Experimental Work in Biology

COMBINED EDITION

D G Mackean

John Murray

This edition © D G Mackean 1983

Individual volumes © D G Mackean 1971 to 1983

This edition first published 1983
by John Murray (Publishers) Ltd
50 Albemarle Street, London W1X 4BD

Reprinted 1988

Printed in Great Britain by Martin's of Berwick

British Library Cataloguing in Publication Data

Mackean, D. G.
  Experimental work in biology.
  Student's book
  1. Biology—Experiments
  I. Title
  574'.07'24   QH316.5
  ISBN 0-7195-4013-5

Contents

General introduction

This book contains a collection of biological experiments which have been tried out by students and found to give reliable results if the instructions are followed exactly and the materials prepared as described in the Teacher's Book.

The experiments do not form a balanced representation of all the topics in an 'O' level or 'CSE' Biology curriculum because not all of these topics lend themselves to experimental work. In some cases, study of plant or animal anatomy is more appropriate to the course; in other cases, an experiment which would be desirable is too complicated or expensive for a school laboratory.

You would not expect to attempt all the experiments in the book. Several of them are alternatives and some may not be appropriate to your syllabus. You will probably try a selection of experiments which are related to certain topics as they crop up during your course.

Whether or not you write an account of each experiment will depend on your teacher, but in all cases it is essential to make a record of the results as they happen. This is why you are usually instructed to draw up a table before starting the experiment or while you are waiting for results.

The results of the experiments are not given, but after each one there is a series of questions. The object of these is to make you think critically about the way the experiment was designed and what you may reasonably deduce from the results. Remember that experimental results cannot be used to *prove* anything. An experiment is a way of testing a theory. If the results are not as expected, the theory is probably no good and may need changing. If the results are just as predicted, the theory stands, but this does not mean it is the only possible theory. More detailed experiments may show it to be inadequate.

Generally speaking, the more results you can collect, the more reliable will be your interpretation. For this reason, it is often a good idea to collect together the results of the whole class rather than base your interpretation on a single experiment.

Acknowledgements

The experiments in this book have been tried out by the students of Sir Frederic Osborn School and modified several times as a result of their experiences and responses to the questions. I am indebted to the laboratory technicians, the students and their teachers, Mrs M. Murphy, Mrs J. Goodfellow and Mr D. Cuckney, whose co-operation and constructive criticism have been invaluable.

I am extremely grateful to Mr Brian Bracegirdle, Ian Mackean and Tony Langham who took the photographs and to the students who posed for them.

It is difficult to identify the origins of ideas for experiments; they arise from so many sources, books, journals, meetings, courses and conversations. Probably none of the experiments in this manual is wholly original and I acknowledge my debt to other authors and teachers. In my search for practical ideas and instructions I have referred to the following:

The Laboratory Manuals and News Letters of the Biological Sciences Curriculum Study.
The Nuffield Biology 'O' level texts and Teachers' guides.
Plants in action by Machlis and Torrey (W. H. Freeman and Co.)
Laboratory Studies in Biology by Lawson, Lewis, Burmester and Hardin. (W. H. Freeman and Co.)
The School Science Review, The Association for Science Education.
Approaches to plant and animal physiology (University of Southampton).
Organisation in Plants by W. M. Baron (Arnold).
Hartshorn, the journal of the Hertfordshire Science Teachers' Association.

1 Food Tests

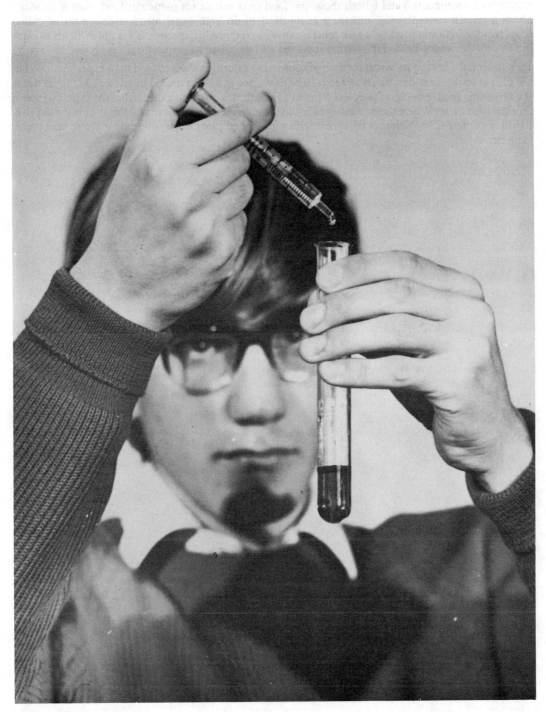

Introduction

It is not expected that all the experiments described in this section will be attempted. Experiments 1, 5, and 10 are basically similar in their aim of establishing that a reagent used in a food test is a more or less reliable indicator of the type of food substance present. Having successfully completed any one of these experiments it would save time to assume that the others will give similar results. This is perhaps not quite so scientific as doing all of the preliminary tests but it is certainly less boring.

Similarly, Experiments 3 and 6 both show that food tests will detect particular foods only if these are above a certain concentration and there is no need to do both experiments unless a set of colour standards for both starch and glucose concentrations is required. Experiment 8 is not difficult to carry out but the reasoning behind it is rather involved and the discussion questions may prove awkward.

Experiment 17 is a simpler visual method of comparing calorific values of food than experiment 16 but the former enables a wider variety of foods to be examined in a short time. Experiment 14 makes use of the preceding food tests (apart from experiment 8) and the instructions assume that the experimenter is familiar with these. Where the quantities of liquid are given in mm (millimetres) this means the depth of liquid in the test-tube. Volumes measured with a graduated pipette or syringe are given as cm^3 (cubic centimetres).

Contents

Experiment 1. The test for starch

(a) Label four test-tubes 1-4.

(b) Put about 20 mm (depth) of
- 10% glucose solution into tube 1
- 1% starch solution into tube 2
- 1% albumen solution into tube 3
- water into tube 4

(c) To each tube, using a dropping pipette, add three drops of iodine solution. Shake the tube (sideways, not up and down) to mix the contents. Look for any colour changes apart from the yellow colour of iodine itself. Copy the table below and record the results in your notebook.

Substance	Colour change after adding iodine
10% glucose solution	
1% starch solution	
1% albumen solution	
Water	

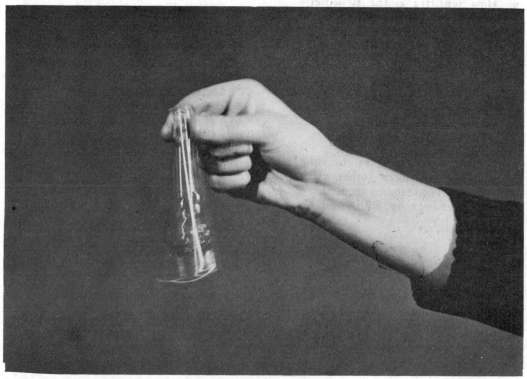

Shake the tube sideways

Experiment 1. Discussion

1   The substances selected for testing are *examples* of three of the principle food materials, sugar (glucose), starch, protein (albumen). With which of these food materials did iodine react to give a colour *change*?

2   Does your result indicate that there is, for example, *no sugar* and *no protein* that will give a colour change with iodine?

3   What experiments would you have to carry out in order to give a confident answer to question 2?

4   Does the result indicate that starch will *always* react with iodine solution to give a colour change? (Try experiment 2.)

5   What was the point of the water in tube 4?

Add three drops of iodine solution to each tube

Experiment 2. Does iodine always give a colour change with starch?

(a) Label five test-tubes 1-5.

(b)　**(i)** Pour about 20 mm (depth) of 1% starch solution into test-tube 1. Using a test-tube holder, heat this solution over a small Bunsen flame until it boils, shaking the tube gently while heating.

　(ii) Add 1 drop of iodine solution.

　(iii) Cool the test-tube under the tap until it feels cold or only slightly warm. If no change occurs on cooling, add another drop of iodine.

(c) Place about 2 mm starch powder in test-tube 2 and add three drops of iodine solution.

(d) Place about 2 mm starch powder in test-tube 3, add about 20 mm cold water and shake vigorously with your thumb over the mouth of the tube. Add one drop of iodine solution.

(e) Pour about 20 mm cold 1% starch solution into test-tube 4 and add one drop of iodine solution.

(f) Pour about 20 mm of cold 1% starch solution into test-tube 5 and add about 5 drops of iodine solution.

Hold the tubes, in pairs, up to the light and shake them gently to compare the colours. Record the results in your notebook as follows:

Substance	Colour change
(b) **(i)** Hot 1% starch solution	
(ii) Cooled 1% starch solution	
(c) Starch powder	
(d) Starch powder in cold water	
(e) 1% starch solution + 1 drop of iodine	
(f) 1% starch solution + 5 drops of iodine	

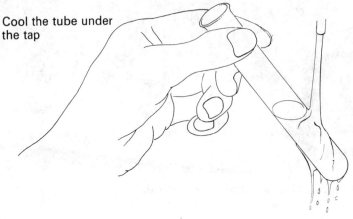

Cool the tube under the tap

Experiment 2.   Discussion

1 Did starch react with iodine to give a colour change in *all* the conditions you tried?
2 When there was a colour change produced, was the colour always the same?
3 What variations in conditions seem likely to influence the colour change?
4 In fact, you are going to use iodine solution as a test for starch in subsequent experiments. Make a general statement (one sentence) which summarizes the visible reaction when iodine solution is added to starch. (You may prefer to try experiment 3 before making a generalization).

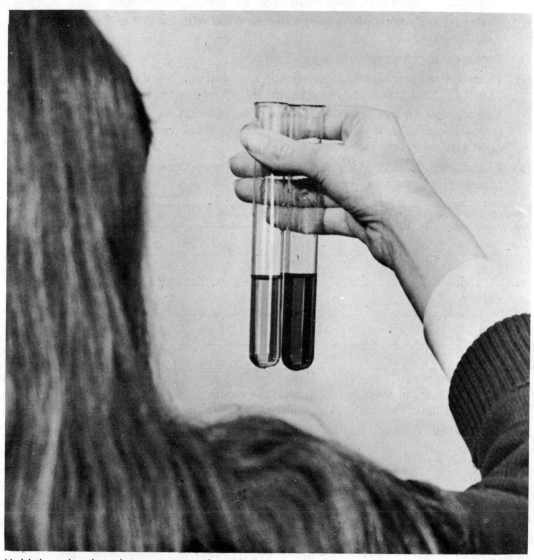

Hold the tubes in pairs to compare the colours

Experiment 3. How sensitive is the starch test?

Make a serial dilution of starch solution as follows:

(a) Place five clean test-tubes in a rack and label them 1-5.

(b) Using a 10 cm^3 pipette, place 10 cm^3 1% starch solution in tube 1.

(c) Take 1 cm^3 of the solution from tube 1 with the pipette and place it in tube 2.

(d) Add 9 cm^3 water with the pipette to tube 2 and shake sideways to mix the contents. The 1 cm^3 of 1% starch solution has now been diluted ten times to make a 0.1% solution.

(e) Pipette 1 cm^3 of this 0.1% starch solution from tube 2 into tube 3.

(f) Add 9 cm^3 water to tube 3 and shake. This will give a .01% solution.

(g) Transfer 1 cm^3 solution from tube 3 to tube 4 and dilute with 9 cm^3 water so making a .001% solution.

(h) Repeat the operation for tube 5.

(i) Tubes 1-4 will now have 9 cm^3 solution in them but tube 5 will have 10 cm^3 so remove 1 cm^3 of this solution from tube 5 so that the subsequent test is a fair one.

(j) Add one drop of iodine solution to each tube and shake to mix. You can add more iodine to each tube, if you think it will make any difference, so long as you add an equal number of drops to each.

(k) Record your results in a table similar to the one below, matching the colours as nearly as possible with crayons.

Strength of solution	Colour after adding iodine	Colour (crayon)
1%		
0.1%		
.01%		
.001%		
.0001%		

NOTE. If you draw the starch solution into your mouth or allow saliva to enter the starch solution it will upset the results. You will have to rinse the pipette and start again with a fresh starch solution.

Experiment 3.   Discussion

1 What is the least concentration of starch that iodine solution will detect?
2 If your answer to question 1 is a single percentage, it will need a qualifying statement.
3 Do you think a .0001% solution contains any starch at all?

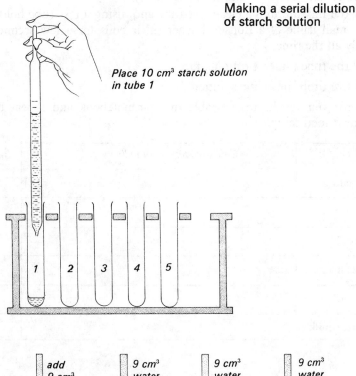

Making a serial dilution
of starch solution

Place 10 cm³ starch solution
in tube 1

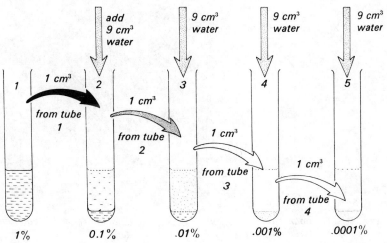

Experiment 4. Testing food for the presence of starch

PRECAUTION. Experiment 3 showed that the starch test is a sensitive one, so if you use a test-tube for more than one experiment, make sure it is *thoroughly cleaned out each time* or else traces of starch from one food sample may remain on the sides and give a positive result for a sample which, in fact, contains no starch. For the same reason, *the mortar and pestle must be washed between each test.*

(a) Label six test-tubes 1-6.
Copy the table given below into your notebook.

(b) Crush a $\frac{1}{2}$ cm cube, or equivalent quantity, of the first food sample in a mortar with about 10 cm^3 water (50 mm in a test-tube).

(c) Pour the mixture into a clean test-tube and, using a test-tube holder, heat the mixture in a small flame of a Bunsen burner till it boils for a few seconds, shaking the tube gently all the time.

(d) Cool the tube under a running tap.

(e) Add five drops of iodine solution.

(f) Record your results in the table in your notebook and repeat the experiment with the next food sample.

	Food sample	Colour change with iodine	Interpretation of result
1	Potato		
2	Onion		
3	Bread		
4	Banana		
5	Apple		
6	Dried milk		

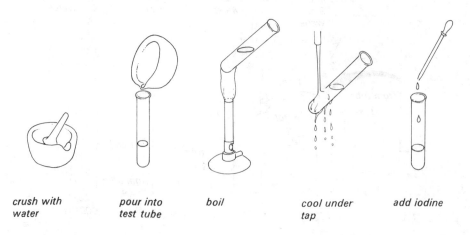

crush with water   pour into test tube   boil   cool under tap   add iodine

Experiment 4. Discussion

1 If a food sample, extracted and tested as above, gives a blue colour with iodine solution, can you assume that starch is present in the sample?

2 Does the result indicate that the only food present is starch?

3 Does the result mean that starch will be present in all samples of the same food?

4 If the food sample, when tested, gives no blue colour, does it mean invariably that no starch is present?

5 What difficulties are there in trying to answer the question, 'Do apples contain starch?'

If you did experiment 3, answer questions 6 and 7.

6 If, with your sample of potato, you obtained a blue colour that corresponded closely to your 1% result in experiment 3, why would you not be justified in saying that 1% of the potato sample was starch? (There are several reasons.)

7 If the blue colour had corresponded to the 0.1% result and the potato sample had been a one centimetre cube, what would you be prepared to say about the concentration of starch in the potato?

What reservations would you make in giving your answer?

Crush the food in a mortar

Experiment 5. The test for sugar

(a) Half fill a beaker with tap water and place it on a tripod and gauze. Heat the water with a Bunsen burner. While waiting for the water to boil, carry on with instructions (**b**) to (**d**).

(b) Label four test-tubes 1-4.

(c) Put 20 mm (depth) of $\left\{\begin{array}{l}\text{1\% starch solution into tube 1}\\\text{10\% glucose solution into tube 2}\\\text{1\% albumen solution into tube 3}\\\text{water into tube 4}\end{array}\right.$

(d) To each tube add about 10 mm Benedict's solution.

(e) Place the test-tubes in the beaker of hot water (Fig. 1), and adjust the flame to keep the water just boiling and then copy the table below into your notebook.

(f) After about 5 minutes, turn out the flame. Place the four tubes in a test-tube rack and compare the colours. Record the results in your table and match the colours as nearly as possible with crayons.

	Solution	Colour change on heating with Benedict's reagent	Final colour (crayon)
1	1% starch		
2	10% glucose		
3	1% albumen		
4	Water		

CLEANING THE TUBES. If a coloured deposit sticks to the inside of the tube even after rinsing with water, it can be removed by pouring in a little dilute hydrochloric acid (HCl). Rinse the tube with water afterwards.

Experiment 5.   Discussion

1 What colour changes took place when Benedict's solution was added to each liquid?

2 The solutions selected for testing are examples of three of the principle food materials, sugar (glucose), starch, protein (albumen). With which of these food materials did Benedict's solution give a decisive change on heating?

3 Apart from the colour, what change took place in the consistency of the Benedict's solution?

4 Do your results indicate that *any* food containing sugar will give the same colour as glucose did when tested with Benedict's solution?

5 Do your results allow you to say that **(a)** no protein will give a colour change when heated with Benedict's solution, **(b)** albumen *never* reacts with Benedict's solution to give a colour change?

6 Can you predict that glucose will always give the same result with Benedict's solution as it did in your experiments? (Try experiment 6.)

7 Why was water included in the test?

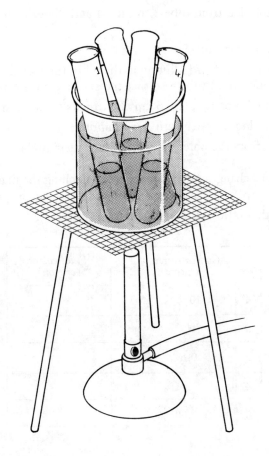

Fig 1. A water bath

Experiment 6.   How sensitive is the Benedict's test?

Make a serial dilution of glucose solution as described below.

(a) Before starting on the serial dilution, prepare a water bath. Half fill a large beaker with water and heat it on a tripod and gauze over a Bunsen burner. Now follow the instructions from **(b)** but watch the water bath and when the water boils turn down the gas to make a small flame, just enough to keep the water at boiling point.

(b) Label five test-tubes 1-5. If sticky labels are used these must be stuck near the top of the tubes.

(c) Using a 10 cm^3 pipette, place 10 cm^3 10% glucose solution in tube 1.

(d) Take 1 cm^3 of this solution from tube 1 with the pipette and place it in tube 2.

(e) Add 9 cm^3 water, with the pipette, to tube 2 and shake sideways to mix the contents. The 1 cm^3 of 10% glucose solution has now been diluted ten times to make a 1% solution.

(f) Pipette 1 cm^3 of this 1% solution of glucose from tube 2 to tube 3.

(g) Add 9 cm^3 water to tube 3 and shake. This will give a 0.1% solution.

(h) Transfer 1 cm^3 solution from tube 3 to tube 4 and dilute with 9 cm^3 water to make a .01% solution.

(i) Repeat the operation for tube 5.

(j) Tubes 1-4 will now have 9 cm^3 solution in them but tube 5 will have 10 cm^3, so remove 1 cm^3 of this solution from tube 5 so that the subsequent test is a fair one.

(k) Using the graduated pipette, place 5 cm^3 Benedict's solution in each tube.

(l) Place all five tubes in the water bath for 3 minutes.

(m) Remove the tubes from the water bath and return them to the test-tube rack to compare their colours.

Tabulate the colour changes in your notebook, matching the final colours with crayons.

CLEANING THE TUBES.   If the coloured deposit remains on the inside of the tube after rinsing it may be removed with a little dilute hydrochloric acid (HCl).

	Concentration of glucose	Final colour (crayon)
1	10%	
2	1%	
3	0.1%	
4	.01%	
5	.001%	

Experiment 6.   Discussion

1 Why was it necessary to use a water bath to heat the test-tubes rather than heat them directly in a Bunsen flame?

2 What is the least concentration of glucose that Benedict's solution will detect? (If your answer to this question is a single percentage, with no qualifying statement, it is likely to be inaccurate.)

3 Do you think a .001% solution of glucose contains any glucose at all?

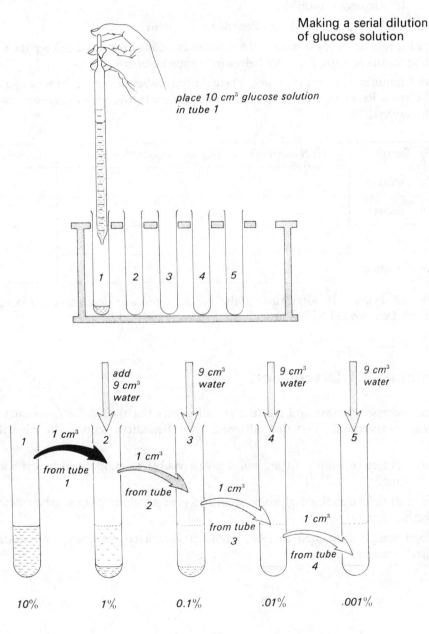

Making a serial dilution
of glucose solution

place 10 cm³ glucose solution
in tube 1

Experiment 7. How specific is Benedict's reagent?

(a) Half fill a beaker with tap water and place it on a tripod and gauze. Heat the water with a Bunsen burner. While waiting for the water to boil, carry on with instructions (b) to (d).

(b) Label four test-tubes 1-4.

(c) Pour about 20 mm (depth) of the following liquids into each tube:
    1.   10% glucose solution
    2.   10% sucrose solution
    3.   10% maltose solution
    4.   10% fructose solution

(d) To each tube add about 10 mm Benedict's solution.

(e) Place the test-tubes in the beaker of hot water and adjust the flame to keep the water just boiling and then copy the table below into your notebook.

(f) After 5 minutes, turn out the flame. Place the four tubes in a test-tube rack and compare the colours. Record the results in your table and match the colours as nearly as possible with crayons.

	Sugar	Final colour after heating with Benedict's solution	Colour (crayon)
1	10% glucose		
2	10% sucrose		
3	10% maltose		
4	10% fructose		

CLEANING THE TUBES.   If, after rinsing the tubes, a reddish deposit still adheres to the inside, it can be removed with a little dilute hydrochloric acid.

Experiment 7.   Discussion

1 Glucose, sucrose, maltose and fructose are all sugars but they differ from each other in chemical composition.   Did they all react with Benedict's solution to give the same result?

2 If a food sample contains a sugar, will it give a visible reaction when boiled with Benedict's solution?

3 If a food sample contained glucose, would it give a colour change when heated with Benedict's reagent?

4 If a food sample contained sucrose, would it give a colour change on heating with Benedict's reagent?

Experiment 8. The test for sucrose

(a) Prepare a water bath by half filling a beaker with water and placing it on a tripod and gauze. Heat the water with a Bunsen flame until it boils and then turn down the flame so that the water just continues to boil.

(b) While waiting for the water to boil, label three test-tubes 1-3 near the rim.

(c) To tube 1 add about 20 mm 10% sucrose solution followed by two drops of dilute hydrochloric acid. Now place tube 1 in the water bath for about two minutes.

(d) In tube 2 place about 20 mm of 10% sucrose solution.

(e) In tube 3 place about 20 mm of water and *two drops only* of dilute hydrochloric acid.

(f) After tube 1 has been in the water bath for 2 minutes, add about 10 mm Benedict's solution to all three tubes.

(g) Place all three tubes in the water bath for one or two minutes.

In your notebook record the results as follows:

	Treatment	Colour change
1	10% sucrose boiled with HCl and tested with Benedict's solution	
2	10% sucrose tested with Benedict's solution	
3	Benedict's solution with hydrochloric acid	

CLEANING THE TUBES. If a red deposit remains on the inside of the tubes after rinsing them out, it can be removed with a little dilute hydrochloric acid.

Experiment 8. Discussion

1 In tube 2, did the sucrose give a colour change on heating with Benedict's solution?

2 In tube 3, did Benedict's solution give any reaction when heated with acid?

3 How was the sucrose solution treated before it would react with Benedict's solution?

4 What was the point of heating Benedict's solution with dilute acid in tube 3?

5 If a food sample contained only sucrose, how would you demonstrate that a sugar was present?

6 If a food sample contained only glucose, what would you observe if you boiled it with dilute acid and then tested it with Benedict's solution?

7 If the only sugar present in a sample of food was sucrose, you would have to carry out two successive tests to prove that *this particular sugar* was present. What are these two tests and why are they both necessary?

Experiment 9.   Testing food samples for the presence of sugar (other than sucrose)

(a)   Prepare a water bath by half filling a beaker with water (hot water from the tap will save time) and heating it on a tripod and gauze over a Bunsen burner. When the water boils, turn the flame down so that boiling point is just maintained.

(b)   While waiting for the water to boil, draw up in your notebook a table like the one below so that the results can be filled in as you obtain them.

(c)   Label six test-tubes 1-6. If you use sticky labels, place these near the tops of the tubes so that they do not float off in the water bath.

(d)   Crush about 1 cm cube, or equivalent volume, of the food in a mortar with about 5 cm³ (20 mm in a test-tube) of water. (If the sample is a liquid, simply pour 20 mm of it into a test-tube.)

(e)   Pour about 20 mm of the crushed mixture into a test-tube and add about 10 mm Benedict's solution.

(f)   Place the test-tube in the water bath for 5-10 minutes and prepare the next food sample.

(g)   Repeat the test with the remaining food samples and compare the colours produced in the test-tubes.

Enter the results in the table in your notebook.

	Food sample	Colour change on heating with Benedict's solution	Interpretation
1	Onion		
2	Milk (or dried milk)		
3	Rice		
4	Raisin (or sultana)		
5	Potato		
6	Banana		

CLEANING THE TEST-TUBES.   If, after rinsing, a film of red copper oxide adheres to the inside of the test-tubes, it can be removed with dilute hydrochloric acid.

Experiment 9.   Discussion

1 If a food sample, extracted and tested as in experiment 9, gives a yellow or red precipitate on heating with Benedict's solution, can you assume that it therefore contains sugar of some kind?

2 Does a yellow or red precipitate indicate that only a sugar is present and not starch or protein?

If you did experiment 6, answer questions 3 and 4.

3 If, with raisins, the colour obtained on testing with Benedict's solution corresponded to the colour you obtained in experiment 6 for 10% glucose, why would you not be justified in saying that raisins contained 10% glucose? (There are several reasons.)

4 (a) What would you need to know and (b) how would you have to design the experiment, to be able to give more precise information about the sugar content of raisins in terms of concentration?

If you did experiment 7 and/or experiment 8 answer questions 5 and 6.

5 Do your results enable you to say whether sucrose is present or absent in any of these food samples?

6 If one of the samples gave no reaction on heating with Benedict's solution, what further tests could you do to see if sucrose was present?

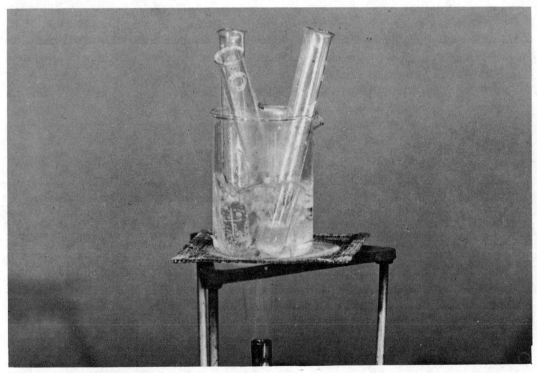

Using a water bath

Experiment 10. The Biuret test

(a) Label four test-tubes 1-4.

(b) Put about 20 mm of
- 1% starch solution into tube 1
- 10% glucose solution into tube 2
- 1% albumen solution into tube 3
- water into tube 4

(c) Pour into each tube, about 5 mm dilute sodium hydroxide. (CARE *)

(d) Add to this about 5 mm dilute copper sulphate solution. Shake the tube sideways to mix the contents.

(e) Return the tubes to the rack, leave for a few seconds and record the resulting colours in your notebook.

	Substance	Reaction with copper sulphate and sodium hydroxide
1	1% starch solution	
2	10% glucose	
3	1% albumen	
4	Water	

* CARE. Sodium hydroxide is caustic and dissolves clothing, skin and bench tops. It is destructive rather than dangerous so if any is spilt on the bench, neutralize it at once with an equal volume of dilute hydrochloric acid and wipe dry. If spilt on clothing do the same but follow with a wash in as much water as possible. If spilt on the skin do not add acid but wash under the tap until the 'soapy' feeling is removed.

Experiment 10. Discussion

1 The substances selected for testing are examples of three of the principal types of food: starch, sugar (glucose), protein (albumen). With which of the samples did the Biuret reaction give a violet colour?

2 Which of these three principle classes of food would you expect to give a violet colour with sodium hydroxide and copper sulphate?

3 Does it follow that all samples of food in this class will give the same reaction with sodium hydroxide and copper sulphate?

4 If a food sample gives a violet colour with sodium hydroxide and copper sulphate, can you say that only the class of food named in 2 (above) is present?

5 What was the point of using test-tube 4 with the water?

Experiment 11.   Testing food samples for protein

(a) In your notebook draw up a table like the one below.

(b) Label six test-tubes 1-6.

(c) If the food is a solid, crush a small piece (about 5 mm cube) in a mortar with about 10 cm³ water. (If the food is a liquid, simply pour about 20 mm of it into a test-tube.)

(d) Pour about 20 mm of the crushed mixture into a test-tube.

(e) Add about 5 mm of sodium hydroxide solution. (CARE *)

(f) Add about 5 mm dilute copper sulphate solution and shake the tube gently sideways to mix the contents. Return the tube to the rack and wait for a few seconds.

(g) Record the results in the table in your notebook.

Food	Result on adding sodium hydroxide and copper sulphate	Interpretation
Milk		
Egg white		
Onion		
Apple		
Beans		
White meat		

* CARE. Sodium hydroxide is caustic and dissolves clothing, skin and bench tops. It is destructive rather than dangerous so if any is spilt on the bench, neutralize it at once with an equal volume of dilute hydrochloric acid and wipe dry. If spilt on clothing, do the same but follow with a wash in as much water as possible. If spilt on the skin do not add acid but wash under the tap until the 'soapy' feeling is removed.

Experiment 11.   Discussion

1 In which of your samples do you think protein is fairly abundant?

2 In which samples do you think protein is absent or in very low concentration?

3 If one of your samples gives no violet colour with the Biuret test, does this result mean that the sample contains no protein?

4 If a food sample gives a violet colour with the Biuret test does this mean that it contains only protein?

Experiment 12.   The test for fats

ALL APPARATUS MUST BE DRY. ALL FLAMES MUST BE EXTINGUISHED.

(a) Label four test-tubes 1-4.

(b) Into tubes 1 and 2 pour about 20 mm alcohol (propan-2-ol).

(c) To tube 1 add one drop of vegetable oil, and shake the tube sideways until the oil dissolves in the alcohol.

(d) In tubes 3 and 4 pour about 20 mm water.

(e) Pour the contents of tube 1 into tube 3 and the contents of tube 2 into tube 4.

(f) Record your results as below:

			Result when added to water
1.	Oil dissolved in alcohol	3.	
2.	Alcohol alone	4.	

CLEANING THE TUBES.   Keep the oily tubes separate from the others and clean them with hot water and liquid detergent.

Experiment 12.   Discussion

1 What was the only difference between the contents of tubes 1 and 2?

2 What was the *visible* difference between tubes 3 and 4 after adding the contents of tubes 1 and 2 to the water in them?

3 How would you attempt to explain the appearance of the liquid in tube 3?

4 What difficulties can you foresee in using this test with food samples which contain both fats (or oils) *and* water?

5 Water and alcohol mix in all proportions. What precautions must be taken in selecting the alcohol for use in this experiment and why is this precaution necessary?

6 How would you design an experiment to find out the sensitivity of this test for fats? (Describe the method in outline only; details of apparatus etc. are not required).

Experiment 13. Testing food samples for fats (the emulsion test)

ALL APPARATUS MUST BE DRY. ALL FLAMES MUST BE EXTINGUISHED.

(a) Label four test-tubes 1-4.

(b) Cut a small sample of food (not more than 5 mm cube).

(c) Crush the sample in a dry mortar and pestle* with about 10 cm³ alcohol (propan-2-ol).

(d) Filter the crushed mixture into a dry test-tube *.

(e) While you are waiting for the first liquid to filter, copy the table given below, into your notebook.
While the second and third samples are filtering, the next food sample can be prepared.

(f) Pour the filtrate into a test-tube containing about 1 cm water.

(g) Examine the liquid in the test-tube and record in your table whether it is clear, cloudy or slightly cloudy.

Food	Appearance of filtrate when added to water	Interpretation
Nuts		
Cheese		
Potato		
Meat		

* The mortar, pestle and filter funnel must be rinsed and carefully dried between each experiment.

Experiment 13.   Discussion

1 Which of the food samples do you consider contains fat (or oil)?

2 If the filtrate was cloudy before adding it to water, does this mean that your results are going to be unreliable?

3 What could be the cause of a cloudy filtrate?

4 If a clear filtrate, when added to water, produces a cloudy liquid, does this indicate that the food sample contained fat (or oil)?

5 If a cloudy filtrate, when added to water, remains the same or goes less cloudy, does it mean that the food sample contained no fat (or oil)?

6 Would it be sufficient to add a crushed food sample directly to water and look for the formation of a cloudy liquid?

Experiment 14. General tests on food samples

In experiments 4, 9, 11 and 13, different kinds of food have been tested, but in each case only one class of food was being sought. The type of investigation most likely to be undertaken by a biologist is one in which a food sample is tested for all possible classes of food.

(a) Draw up a table like the one below, in your notebook, for each sample of food you propose to test.

(b) Divide the food sample into two unequal portions, approximately 3:1 in volume.

EXTINGUISH ANY BUNSEN FLAME.

(c) Crush the smaller portion of food in a dry mortar and pestle with about 10 cm³ alcohol (propan-2-ol). Filter into a dry test-tube and pour half the filtrate into about 10 mm water in a test-tube. Record your observations. (The test for fats is described fully on p. 27.) Dispose of the filter paper in the receptacle provided.

(d) Return to the larger portion of food and crush it in a mortar and pestle with about 20 cm³ water (about 100 mm in a test-tube).

(e) Label three test-tubes 1-3 and share the crushed mixture between them so that each contains about 20 mm of liquid.

(f) See that no alcohol remains on your bench, then light the Bunsen burner and use it to heat a water bath.

(g) To tube 1 add about 10 mm Benedict's solution and heat it in the water bath for about 5 minutes. (For full details of this test see p. 22.) Record your observations.

(h) To tube 2 add about 5 mm sodium hydroxide solution and 5 mm dilute copper sulphate solution. Shake the tube from side to side. (Full details on p. 25.) Record your observations.

(i) Boil the contents of tube 3. Cool under the tap and add five drops of iodine solution. (Full details on p. 14.) Record your observations.

Food sample		
Test	*Result*	*Interpretation*
Extracted in alcohol and poured into water		
Benedict's solution		
Biuret reaction		
Iodine		

Experiment 14. Discussion

1 For what reasons do you think a biologist would want this information?
2 To what use could he put the information?
3 What other facts about food samples is he likely to need to know?

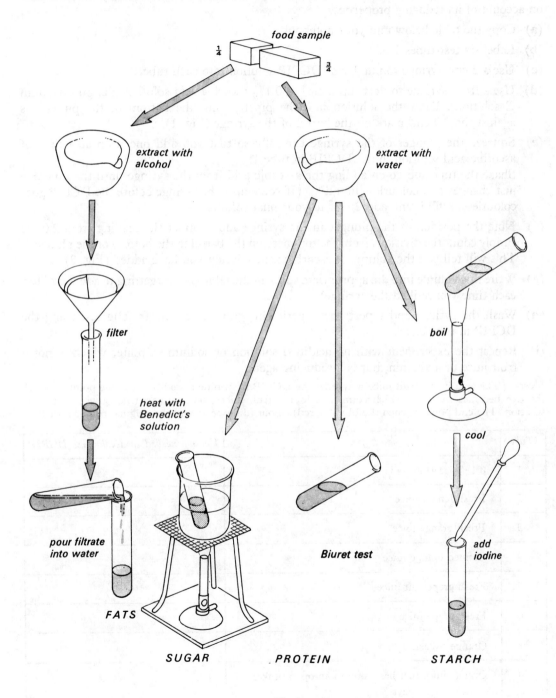

food sample

¼   ¾

extract with alcohol

extract with water

filter

boil

heat with Benedict's solution

cool

pour filtrate into water

Biuret test

add iodine

FATS

SUGAR   PROTEIN   STARCH

Experiment 15.   A comparison of the vitamin C content of fruit juices

Vitamin C is *ascorbic acid:* one of its chemical properties is that it is a 'reducing agent', i.e. removes oxygen from or adds hydrogen to other chemicals.

DCPIP is dichlorophenol-indophenol, a blue dye which is decolourized by ascorbic acid on account of its reducing properties.

(a) Copy the table below into your notebook.

(b) Label six test-tubes 1-6.

(c) Use a 2 cm^3 syringe to put 1 cm^3 DCPIP solution into each tube.

(d) Use a fresh syringe to draw up 2 cm^3 of 0.1% ascorbic acid solution (i.e. pure vitamin C solution). Draw the solution into the pipette until the bottom of the plunger is against the 2.0 cm^3 mark on the barrel of the syringe (Fig. 1).

(e) Squeeze the plunger of the syringe carefully so that you add one drop at a time of ascorbic acid solution to the DCPIP in tube 1.
Shake the tube and go on adding the ascorbic acid from the syringe until the blue dye just changes to a colourless liquid*.   (If you empty the syringe before the DCPIP goes colourless, refill it and add 2 cm^3 to your final volume.)

(f) Note the position of the plunger in the syringe and subtract the reading from 2.0 (or simply count the divisions from the top mark on the barrel to the bottom of the plunger). This will tell you the volume of ascorbic acid solution you have added (Fig. 2).

(g) Write this volume into the appropriate space in the table (not forgetting to add 2 cm^3 for each time you refilled the syringe).

(h) Wash the syringe and repeat the experiment with one of the fruit juices, using the DCPIP in tube 2.

(i) Repeat the experiment with an acidified solution of sodium sulphite, which is not a fruit juice or a vitamin, but is a reducing agent.

Note.   The acid in the fruit juice may turn the DCPIP from blue to red but it is the point at which the dye becomes *colourless* which counts. Coloured fruit juices, however, will not give a colourless solution. The end point is when the blue (or red) colour disappears and the original colour returns.

Tube	Liquid	Volume needed to decolourize DCPIP
1	0.1% ascorbic acid	
2	Fresh lemon juice	
3	Fresh orange juice	
4	Canned orange juice	
5	Fresh grapefruit juice	
6	Fresh grape juice	
7	Orange squash	
9	Orange juice that has stood in an open beaker for .... days	

Experiment 15. Discussion

1 If the volume of fruit juice needed to decolourize DCPIP is greater than the volume of 0.1% ascorbic acid, does this mean that the juice contains more or less vitamin C than ascorbic acid solution?

2 On this basis, which of your samples contained the most vitamin C?

3 Did the canned fruit juice differ greatly from the fresh juice?

4 Suppose that 0.8 cm³ orange juice decolourized 1 cm³ DCPIP but it took 1.6 cm³ grapefruit juice to decolourize the same volume of DCPIP; does this mean that oranges contain more vitamin C than grapefruit?

5 What might take place at a canning factory which could (a) reduce, (b) increase the vitamin C content of fruit juice?

6 Did the fruit juice which had been exposed to air for several days have significantly less vitamin C than the fresh juice?

7 Sodium sulphite is not vitamin C and yet it decolourized DCPIP. Why did it do this? (See introductory notes on p. 30 and instruction (i).)

8 Since chemicals other than vitamin C will decolourize DCPIP, does this mean that using the dye as a test for vitamin C is unsatisfactory?

9 If canned or bottled juice has sulphur dioxide or sodium sulphite added as a preservative, will this affect the reliability of your results with DCPIP?

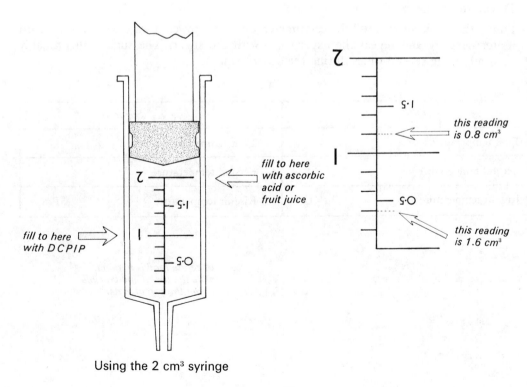

Using the 2 cm³ syringe

Experiment 16. A comparison of the energy values of oil and sucrose

(a) Before starting the experiment, copy the tables given below into your notebook.

(b) You are provided with two crucibles. One contains 1 g sugar and the other contains 1 g cooking oil.

(c) Place the crucible with the oil in a triangle arranged on a tripod.

(d) Fill the calorimeter (a metal can) with 100 cm³ cold water, using a measuring cylinder. Clamp the can, as shown in the diagram, so that the tripod and crucible can be slid under it, leaving a gap of about 50 mm.

(e) Measure and record the temperature of the water (1st temperature).

READ INSTRUCTIONS **(f)**, **(g)** and **(h)** before igniting the oil.

(f) Keeping the tripod and crucible clear of the calorimeter, heat the crucible gently from below with the Bunsen flame. Hold the Bunsen burner in your hand and from time to time bring the flame to the mouth of the crucible until the oil just catches alight. As soon as this occurs, slide the tripod and crucible under the calorimeter so that the flame heats the bottom of the can.

(g) If necessary, adjust the height of the calorimeter so that the flame either **(i)** is not too far from the can or **(ii)** does not spread round the sides too much.
Relight the oil if it goes out. Do this by sliding the tripod and crucible over the Bunsen and *not* by putting the Bunsen under the crucible while it is beneath the calorimeter.

(h) When all the oil has burned away, measure the temperature of the water in the calorimeter and record it in your notebook (2nd temperature).
Do not touch the crucible until it is cool.

(i) Empty the water away; refill the calorimeter with 100 cm³ cold water as before; record its temperature and repeat the experiment with the sugar. The sugar will probably need relighting several times during the experiment.

RESULTS

Oil			Sugar		
First temperature	=		First temperature	=	
Second temperature	=		Second temperature	=	
Rise in temperature	=	(T °C)	Rise in temperature	=	(T °C)

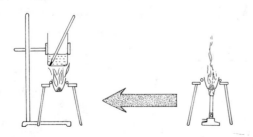

*once the food is burning,
slide the tripod and crucible
under the can of water*

CALCULATION

Copy the calculation given below into your notebook, inserting your own figures for the temperature rise T.

Oil

1 g water heated 1 °C needs 1 cal

∴ 100 g water heated 1 °C needs 1 × 100 cals

∴ 100 g water heated T °C needs 1 × 100 × T cals ≡   cals

∴ 1 g oil gave   cals

[1 000 cals ≡ 1 kilo-calorie (kcal)]

∴ 1 g oil gave $\dfrac{}{1000}$ kcals ≡   kcals per gram

Sugar

Repeat the calculation for sugar and enter the results in a table like the one on the right.

Food	kcals per g
Oil	
Sugar	

Experiment 16. Discussion

The idea of the experiment is that when 1 g food is burned the heat is transferred to the water. From the rise in temperature of the water can be calculated the number of calories given out by the burning food. The greater the rise in temperature the more will be the calorific value of the food.

1 Why was so much emphasis placed on keeping the Bunsen flame clear of the calorimeter?

2 What are the sources of innacuracy in assuming that the heat is transferred from the burning food to the water?

3 What kind of residue was left after burning (a) the oil, (b) the sugar?
What do you think is the chemical nature of the residue from the burnt sugar and how does it contribute to the sources of error in this experiment?

4 Will the inaccuracies discussed in 2 and 3 make the results too high or too low?

5 Although this experiment is a very crude one, it is still reasonable to use the results to *compare* the energy obtained from sugar and fat. Why can the sources of error be ignored for this comparison?

6 Which of the samples released more energy during burning?

7 Is it reasonable to suppose that this food will also provide more energy when it is eaten and used in the body? Justify your answer.

Experiment 17. Comparison of energy values of food

READ ALL THE INSTRUCTIONS BEFORE STARTING.

(a) Place a little sugar in the bottom of a dry test-tube.

(b) Using a test-tube holder, heat the bottom of the tube strongly in a Bunsen flame.

(c) Look for steam coming off or water condensing on the cooler upper part of the test-tube.

(d) When smoke starts to come out of the tube, try to set it alight with a lighted splint. Keep heating the tube and go on trying to light the smoke until no more appears.

(e) DO NOT PUT THE TUBE ON THE BENCH OR IN THE RACK but place it on a heat-resistant mat until cool.

(f) Draw up the table below in your notebook.

(g) Repeat this test with different food samples and record your results.

Food	How much steam or condensed water?	How much smoke?	Did it burn vigorously, briefly or not at all?
Sugar			
Apple			
Cheese			
Potato			
Nuts			
Dried potato			

Experiment 17. Discussion

1   Is it likely that what happens to food in the test-tube is similar to what happens in the body? Give reasons.

2   In this experiment, in what forms is the energy released from the food?

3   In the body, in what forms is energy released?

4   From the experiment, what kind of result would you regard as showing a great deal of energy present in a food sample?

5   On this basis, which food samples contained most energy?

6   Does it follow that the foods listed in the answer to 5 will supply most energy to the body? Explain.

7   When a food sample, on heating, fails to produce any inflammable gas, does this signify that it contains little energy? Explain.

8   Suppose a food sample *is* rich in energy and *does* give an inflammable gas on heating but the food also contains a great deal of water. What problems can you see in conducting and interpreting this experiment?

9   What is the point of eating food which contains very little energy?

2 Enzymes

Introduction

Eleven of the experiments in this section are concerned with digestive enzymes. This is because their activity is readily understood and many of them are easily and cheaply obtained. It must not be thought, however, that enzymes are exclusively concerned with digestion or even with breaking down complex molecules to simpler molecules. Of the hundreds of reactions taking place in the cytoplasm of a cell, each one needs a particular enzyme to make it work effectively and there are at least as many 'building-up' reactions as there are 'breaking-down' reactions. Experiments 8, 10 and 11 deal with enzymes which occur in living cells and are not concerned with digestion. Experiment 8 shows an enzyme building a complex molecule from relatively simple ones.

Some of the alternative experiments, e.g. 2a and 3a, offer simpler experimental situations while others merely use a different enzyme or substrate. It is not suggested that *all* the experiments should be attempted, at least not within a short period of time. It may be advisable to return to this section from time to time and conduct an experiment that has been omitted on a previous occasion.

In order to write experiments whose outcome can be predicted with reasonable certainty, it has been necessary to specify exact concentrations and volumes of the solutions used. Consequently, it is essential that the experimenter should become reasonably adept at using a graduated pipette for measuring small volumes of liquids.

Where quantities of liquid are given in mm (millimetres) this means the depth of liquid in the test-tube. Volumes measured with a graduated pipette are given as cm^3 (cubic centimetres).

Uses of saliva

There is no evidence at the moment that the AIDS virus can be transmitted by saliva. However, students should wash their own glassware if it has contained their saliva. If it seems preferable to avoid use of saliva, 5% diastase or aminase will give similar results.

Contents

Experiment 1. The hydrolysis of starch with hydrochloric acid

(a) Prepare a water bath by half filling a 250 cm³ beaker with warm water and heating it to boiling point on a tripod and gauze, with a Bunsen burner. When the water boils, reduce the flame to keep the water at boiling point.

(b) Label four test-tubes 1-4.

(c) Copy the table given below into your notebook.

(d) In each tube place 5 cm³ 3% starch solution.

(e) Using a graduated pipette, add 3 cm³ Benedict's solution to the starch solution in tube 1 and place the tube in the boiling water bath for five minutes.

(f) Rinse the graduated pipette and use it to add 1 cm³ dilute hydrochloric acid to the starch solution in each of tubes 2, 3 and 4. *Note the time* and place all three tubes in the water bath. (They will be removed at five, ten and fifteen minutes respectively).

(g) Remember to remove tube 1 from the water bath after five minutes if you have not already done so.

(h) After five minutes, remove tube 2 from the water bath and cool it under the tap. Neutralize the acid by adding solid sodium bicarbonate, a little at a time, until the addition of one portion produces no fizzing. Place tube in the rack and return to tube 3.

(i) After ten minutes in the water bath, remove tube 3, cool and neutralize the contents as described in (h). Place the tube in the rack.

(j) After fifteen minutes in the water bath, remove tube 4; cool and neutralize as before, and place it in the rack.

(k) With a dropping pipette, remove a sample of the liquid from tube 2 and place 3 drops on a spotting tile. Rinse the pipette and repeat the procedure for tubes 3 and 4. Add one drop of dilute iodine to each drop of liquid on the tile and record the colour.

(l) Rinse the graduated pipette and use it to place 3 cm³ Benedict's solution in each of tubes 2, 3 and 4. Return all three tubes to the water bath and heat for five minutes. After this time replace the tubes in the rack and allow them to cool sufficiently to handle. Hold the tubes to the light to compare the colours of the solutions and compare also the colours and relative quantities of any precipitates. Record the colours.

Tube	Containing	Appearance with iodine	Appearance after testing with Benedict's		
			Colour of solution	Colour of precipitate	Relative quantity of precipitate
1	3% starch solution only	————			
2	3% starch solution boiled for 5 min with dilute HCl				
3	3% starch solution boiled for 10 min with dilute HCl				
4	3% starch solution boiled for 15 min with dilute HCl				

Experiment 1. Discussion

1 What was the point of adding sodium bicarbonate to tubes 2, 3 and 4?

2 What food substance is Benedict's solution a test for?

3 At the end of the experiment, what food substance was present in tubes 3 and 4 that was not there at the beginning?

4 What evidence have you that this substance was not present at the beginning of the experiment?

5 How do you account for the difference, after testing with Benedict's solution, between tubes 2, 3 and 4?

6 How do you interpret the results of the iodine test in tubes 2, 3 and 4?

7 What relationship is there between the interpretation of the results with the iodine test and the Benedict's test?

8 The starch molecule consists of a long chain of carbon atoms with hydrogen and oxygen atoms attached. Sugars, such as glucose, consist of six carbon atoms with hydrogen and **oxygen atoms attached.**

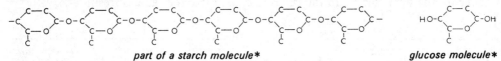

part of a starch molecule∗   *glucose molecule∗*

[∗*many H atoms omitted*]

Assuming that the hydrochloric acid is acting only as a catalyst in the reaction, attempt an explanation of the chemical change which takes place in tubes 3 and 4.

9 In this experiment, the emphasis is on the conversion of starch to something else using hydrochloric acid. What control experiment would have to be carried out to show that hydrochloric acid played a significant part in bringing about this change?

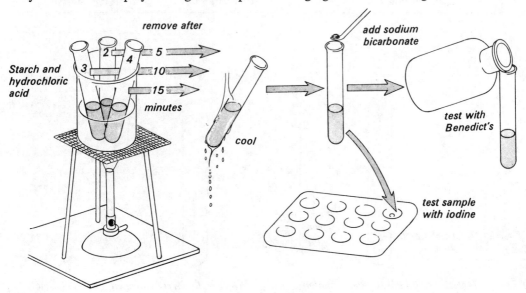

Experiment 2a.   The action of saliva on starch

(a) Prepare a water bath by using a Bunsen burner to heat some water in a beaker on a tripod and gauze till it boils; then turn down the flame to keep the water just boiling. While waiting for the water to boil carry on from **(b)**.

(b) Label five test-tubes 1-5 and in tube 5 collect saliva as follows:
    (i) Rinse the mouth with water to remove food residues.
    (ii) Chew paraffin wax to stimulate the flow of saliva.
    (iii) Collect about 30 mm (depth) saliva.

(c) Using a graduated pipette, place 5 cm³ 2% starch solution in tubes 1-4.

(d) Rinse the pipette and draw up 4 cm³ saliva. Place 2 cm³ saliva in each of tubes 2 and 3 and shake the tubes to mix the contents. Leave the tubes for 5 minutes and copy the table below into your notebook.

(e) After 5 minutes, add 3 drops of iodine to tubes 1 and 2.

(f) Use the graduated pipette to add 3 cm³ Benedict's solution to tubes 3 and 4 and place both tubes in the water bath for 5 minutes.

(g) Compare the final colours in the tubes and complete the table of results.

Tube	Contents	Tested with	Result	Interpretation
1	2% starch solution	iodine		
2	2% starch solution + saliva	iodine		
3	2% starch solution + saliva	Benedict's		
4	2% starch solution	Benedict's		

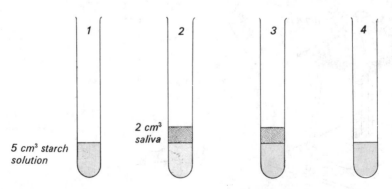

5 cm³ starch solution

2 cm³ saliva

AFTER 5 MINS—add 3 drops iodine solution　　　　*AFTER 5 MINS—test with Benedict's solution*

40

Experiment 2a. Discussion

1 What normally happens when iodine solution is added to starch?
2 Tube 2 contained starch solution at the beginning of the experiment; how do you explain the reaction with iodine at the end of the experiment?
3 What food substance is Benedict's solution a test for?
4 Was this food substance present in tubes 3 or 4 at the *beginning* of the experiment?
5 What evidence have you to support your answer?
6 What evidence is there to suggest that this food substance is present in tube 3 at the *end* of the experiment?
7 What chemical change could have taken place in tubes 2 and 3 after adding saliva, which would explain the results in these tubes after applying the iodine test and Benedict's test?
8 What part could saliva have played in this chemical change?
9 Suggest a control to the experiment which would help to support your answer to 7.

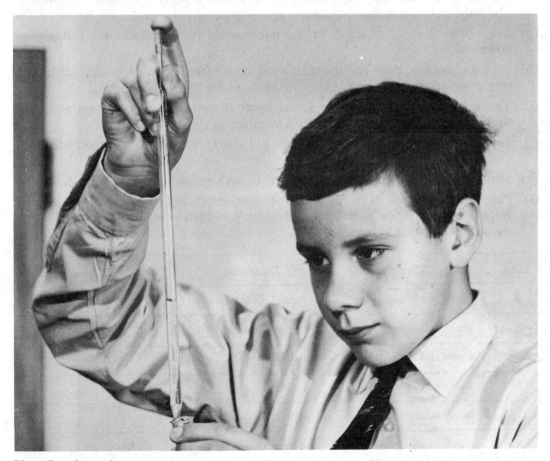

Place 5 cm³ starch solution in each tube

Experiment 2b. The action of saliva on starch

Study the flow chart on p. 43 for a few minutes to gain an idea of the outline of the experiment.

(a) Prepare a water bath by using a Bunsen burner to heat some water in a beaker on a tripod and gauze till it boils; then turn the flame down to keep the water just boiling. While waiting for the water to boil, carry on from **(b)**.

(b) Label eight test-tubes 1-8 and in tube 1 collect saliva as follows:

    (i) Thoroughly rinse the mouth with water to remove food residues.
    (ii) Chew paraffin wax to stimulate the flow of saliva.
    (iii) Collect about 50 mm (depth) saliva.

(c) Pour half the saliva into tube 2 and place the tube in the boiling water bath for 3 minutes.

(d) Using a graduated pipette, add 5 cm^3 2% starch solution to tubes 3, 4 and 7.

(e) Rinse the pipette and use it to transfer 5 cm^3 boiled saliva from tube 2 to tube 3. Shake the tube sideways to mix the contents.

(f) Use the graduated pipette to transfer 5 cm^3 unboiled saliva from tube 1 to tube 4. Shake the tube to mix the contents.

(g) Leave tubes 3 and 4 to stand for five minutes and copy the table below into your notebook.

(h) After five minutes, pour half the contents of tube 3 (the boiled saliva and starch) into tube 5 and add three drops of iodine solution to tube 5.

(i) To the remaining liquid in tube 3, add about 20 mm Benedict's solution and place the tube in the boiling water bath for 5 minutes.

(j) Pour half the contents of tube 4 (starch and saliva) into tube 6 and then add three drops of iodine to tube 6.

(k) Test the remaining liquid in tube 4 with Benedict's solution as you did in **(i)**.

(l) Pour half the contents of tube 7 (starch solution) into tube 8 and test the two samples respectively with iodine as in **(h)** and Benedict's solution as in **(i)**.
Record the results in your table.

Tube	Contents	Tested with	Result	Interpretation
3	starch and boiled saliva	Benedict's solution		
4	starch and saliva	Benedict's solution		
5	starch and boiled saliva	iodine		
6	starch and saliva	iodine		
7	starch solution (control)	iodine		
8	starch solution (control)	Benedict's solution		

Experiment 2b. Discussion

1 What substances do iodine and Benedict's solution test for?

2 What change takes place when starch and saliva are mixed, according to the results in tubes 4 and 6?

3 Tubes 3 and 5 probably did not give the same results as tubes 4 and 6. In what way were the contents treated that could account for this difference?

4 (a) Are your results consistent with the hypothesis (theory) that an enzyme in saliva has changed starch to sugar?
   (b) Do your results prove that an enzyme in saliva has changed starch to sugar?

5 In what way do the results with tubes 3 and 5 support the enzyme hypothesis?

6 Do your experimental results rule out the possibility that (a) starch converts unboiled saliva to sugar or (b) starch and unboiled saliva combine chemically to form sugar?

7 The starch molecule consists of a long chain of carbon atoms with oxygen and hydrogen atoms attached. A sugar, such as glucose, has molecules consisting of six carbon atoms with oxygen and hydrogen atoms attached (see p. 39). Using this information, suggest a way in which sugar could be formed from starch. What part would an enzyme play in this reaction?

8 If you tried experiment 1, state in what ways the conditions for the reaction between starch and hydrochloric acid differed from those in the starch/saliva reaction.

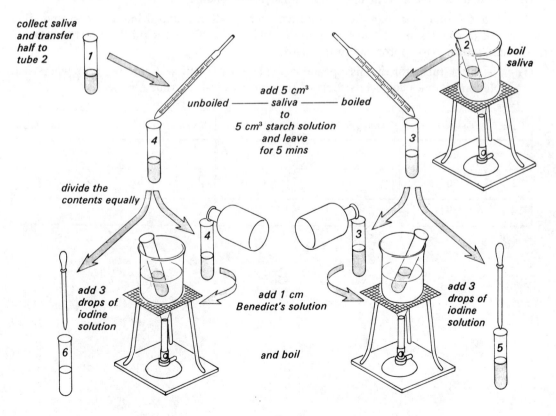

Experiment 3a. The effect of temperature on enzyme activity

(a) Label seven test-tubes 1-7.

(b) Collect saliva in tube 7 as follows:
   (i) Rinse the mouth with water to remove food residues.
   (ii) Chew paraffin wax to promote the flow of saliva.
   (iii) Collect about 20 mm saliva in tube 7.

(c) Using a graduated pipette, place 5 cm³ 1% starch solution in tubes 1-3.

(d) To each of these tubes add also 6 drops of *dilute* iodine solution.

(e) Wash the graduated pipette and use it to place 1 cm³ saliva in each of tubes 4-6.

(f) Prepare three water baths by half filling 250 cm³ beakers or jars as follows:
   (i) ice and water, adding ice during the experiment to keep the temperature at about 10 °C.
   (ii) water from cold tap at room temperature, about 20 °C.
   (iii) warm water at about 35 °C by mixing hot and cold water from the tap.

(g) Place tubes 1 and 4 in the cold water, 2 and 5 in the water at room temperature and 3 and 6 in the warm water. Leave the apparatus for five minutes so that the saliva and starch solution attain the temperature of the water bath.

(h) Draw up a table in your notebook similar to the one below.

(i) Note the time and then pour the saliva from tubes 4, 5 and 6 into the corresponding tube of starch solution, starting with the coldest one. Shake the tube to mix the contents and return it to appropriate water bath.

(j) Watch the tubes for the disappearance of the blue colour and note the time when each solution becomes colourless. Record the time intervals in your notebook.

Tube	Temperature	Time for blue colour to disappear	Speed of reaction (fast, medium, slow)
1	10 °C		
2	20 °C		
3	35 °C		

Experiment 3a. Discussion

1 What is the significance of the disappearance of the blue colour from the starch/iodide mixture?

2 At what temperature did the reaction proceed most rapidly?

3 Make a general statement connecting the rate at which saliva hydrolyses starch and the temperature at which it is acting. Put in any qualifications you think necessary.

4 Would you expect to find any upper limit of temperature beyond which the reaction ceases to get any faster?

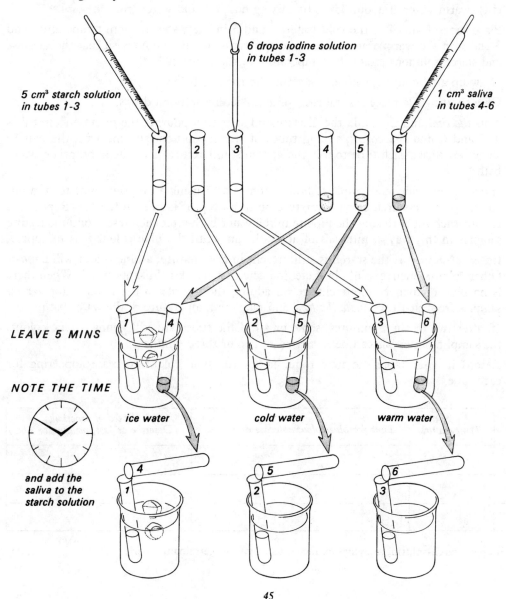

5 cm³ starch solution in tubes 1-3

6 drops iodine solution in tubes 1-3

1 cm³ saliva in tubes 4-6

LEAVE 5 MINS

NOTE THE TIME

ice water   cold water   warm water

and add the saliva to the starch solution

Experiment 3b. The effect of temperature on enzyme activity

(a) Label six test-tubes 1-6.

(b) Using a graduated pipette, place 5 cm³ 1% starch solution in tubes 1-3.

(c) Wash the pipette and use it to place 5 cm³ 5% diastase* solution in tubes 4-6.

(d) Prepare three water baths by half filling 250 cm³ beakers as follows:

    (i) ice and water, adding ice during the experiment to keep the temperature at about 10 °C.

    (ii) water from the cold tap at room temperature, about 20 °C.

    (iii) warm water at about 35 °C by mixing hot and cold water from the tap.

(e) Place tubes 1 and 4 in the cold water, 2 and 5 in the water at room temperature and 3 and 6 in the warm water. Leave the apparatus for five minutes so that the diastase and starch solutions attain the temperature of the water bath.

(f) Draw up a table in your notebook similar to the one below.

(g) On a spotting tile place several rows of *dilute* iodine solution drops.

(h) *Note the time* and then mix the diastase and starch by pouring the diastase from tubes 4, 5 and 6 into the corresponding tubes of starch solution, starting with the pair in ice water. Shake each tube to mix the contents and return it to the appropriate water bath.

(i) Immediately remove a sample of liquid from tube 3 using a dropping pipette, and put one drop of the starch/enzyme mixture on to a drop of iodine on the tile. Repeat the test for each tube, rinsing the pipette in the water bath after each test. Continue testing samples in this way at intervals of about 1 min. until the blue colour fails to appear.

(j) *Colour changes.* If the starch is still present in the mixture, a blue colour will appear. Other colours, mauve, pink, buff etc. indicate that starch is 'disappearing'. When there is no blue or mauve colour change on adding the sample to the iodine, stop taking samples from that tube, *note the time* and take the temperature of the water bath.

(k) If, after two or three minutes, some or all of the tubes are still giving a blue colour, the sampling intervals can be increased to two or three minutes.

Record in your table the time required for the blue colour to cease appearing for each tube.

Tube	Temperature	Time for blue colour to cease appearing	Speed of reaction. (Describe as fast, medium or slow)
1			
2			
3			

* A group of starch-splitting enzymes extracted usually from germinating barley.

Experiment 3b. Discussion

1 What is the significance of the eventual failure of the starch/enzyme mixture to give a blue colour with iodine?

2 At what temperature *did* the reaction proceed most rapidly?

3 What temperature, do you think, is the most favourable for the reaction?

4 How would you have to modify the experiment to answer question 3 satisfactorily?

5 In the experiment, does it matter that the size of the sample (the size of the drop of starch/diastase) from the test-tube is likely to vary?

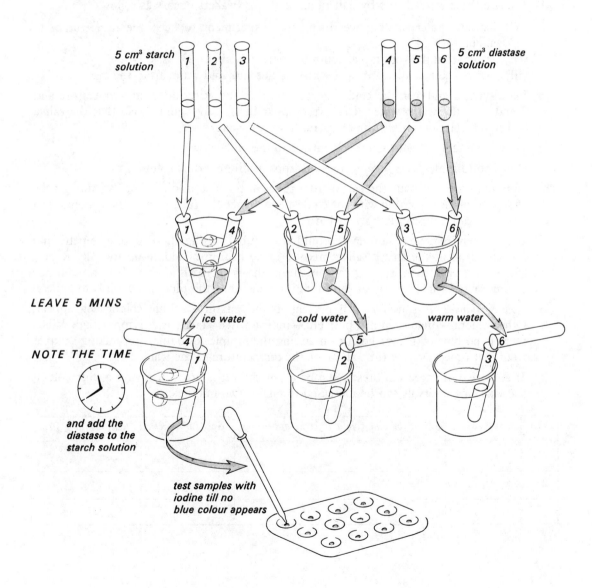

Experiment 3c. The effect of temperature on enzyme activity

(a) Label seven test-tubes 1-7 and collect saliva in tube 7 as follows:
  (i) Rinse the mouth with water to remove food residues.
  (ii) Chew paraffin wax to promote the flow of saliva.
  (iii) Collect about 20 mm saliva in test-tube 7.

(b) Using a graduated pipette, place 10 cm³ 2% starch solution in each of tubes 1-3.

(c) Wash the pipette and use it to place 1 cm³ saliva in each of tubes 4-6.

(d) Prepare three water baths by half filling 250 cm³ beakers or jars as follows:
  (i) ice and water, adding ice during the experiment to keep the temperature at about 10 °C.
  (ii) water from the cold tap at room temperature, about 20 °C.
  (iii) warm water at about 35 °C by mixing hot and cold water from the tap.

(e) Place tubes 1 and 4 in the cold water, 2 and 5 in the water at room temperature and 3 and 6 in the warm water. Leave the apparatus for five minutes so that the saliva and starch solutions attain the temperature of the water bath.

(f) Draw up a table in your notebook similar to the one below.

(g) On a spotting tile place several rows of drops of dilute iodine solution.

(h) Note the time and pour the saliva from tubes 4, 5 and 6 into the corresponding tube of starch solution, starting with the coldest one. Shake the tube to mix the contents and return it to the appropriate water bath.

(i) Immediately remove a sample of liquid from tube 3 using the dropping pipette and put one drop of the starch/saliva mixture on to a drop of iodine on the tile. Repeat the test for each tube, rinsing the pipette in the water bath after each test. Continue testing samples in this way at ½ minute intervals until the blue colour fails to appear.

(j) *Colour changes*. If starch is still present in the mixture, a blue colour will appear. Other colours—mauve, pink, buff etc.—indicate that starch is 'disappearing'. When there is no blue or mauve colour on adding the sample to iodine, stop taking samples from that tube, note the time and take the temperature of the water bath.

(k) If after two or three minutes, some or all of the tubes are still giving a blue colour, the sampling intervals can be increased to one or two minutes.

Tube	Temperature	Time for blue colour to cease appearing	Speed of reaction (fast, medium, slow)
1			
2			
3			

Experiment 3c.   Discussion

1 What is the significance of the eventual failure of the starch/enzyme mixture to give a blue colour with iodine?

2 At what temperature *did* the reaction proceed most rapidly?

3 What temperature, do you think, is most favourable for the reaction?

4 How would you have to modify the experiment to answer question 3 satisfactorily?

5 In the experiment, does it matter that the size of the starch/saliva sample (the drop size) is likely to vary?

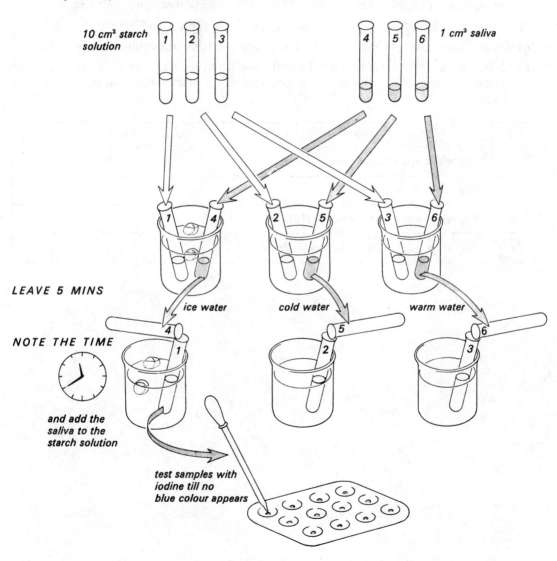

10 cm³ starch solution

1 cm³ saliva

LEAVE 5 MINS

ice water   cold water   warm water

NOTE THE TIME

and add the saliva to the starch solution

test samples with iodine till no blue colour appears

Experiment 4. The effect of pepsin on egg-white suspension

(a) Label four test-tubes 1-4.

(b) Into each tube place about 5 cm^3 (20 mm in test-tube) egg-white suspension.

(c) Use a dropping pipette to add three drops of dilute hydrochloric acid to tubes 2, 3 and 4.

(d) Using a graduated pipette or syringe place 1 cm^3 1% pepsin solution in a clean test-tube and heat it over a small Bunsen flame until the liquid boils. Add the boiled pepsin to the egg-white suspension in tube 4.

(e) Prepare a water bath in a 250 cm^3 beaker or jar by mixing hot and cold water from the tap to attain a temperature of about 40 °C. Have the beaker about half full.

(f) Using a graduated pipette, add 1 cm^3 1% pepsin to tubes 1 and 3 only.

(g) Place all four tubes in the water bath and copy the table below into your notebook.

(h) After five or six minutes remove the four tubes from the water bath and replace them in the test-tube rack. Compare the appearance of the contents and fill in your table of results.

Tube	Contents	Result
1	Egg-white suspension & pepsin	
2	Egg-white suspension & HCl	
3	Egg-white suspension, pepsin & HCl	
4	Egg-white suspension, boiled pepsin & HCl	

Experiment 4.   Discussion

1 Why, do you suppose, does the egg-white suspension used in the experiment look cloudy?

2 If the egg-white suspension goes from cloudy to clear, what change must have occurred?

3 In which of the test-tubes are the conditions most like those in the stomach?

4 In general, how is an enzyme affected by boiling?

5 Do the results with tubes 3 and 4 prove that pepsin is an enzyme?

6 Are the results with tubes 3 and 4 consistent with the theory that pepsin is an enzyme that can accelerate the digestion of egg-white?

7 Are the results consistent with the theory that pepsin is an enzyme which, in acid conditions, can digest proteins?

8 Suppose the hypothesis is advanced that hydrochloric acid is an enzyme but can digest egg-white only in the presence of unboiled pepsin, what control experiment would help to eliminate this explanation?

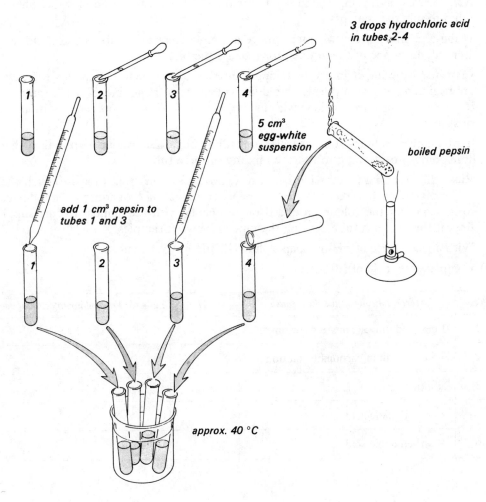

3 drops hydrochloric acid in tubes 2-4

5 cm³ egg-white suspension

boiled pepsin

add 1 cm³ pepsin to tubes 1 and 3

approx. 40 °C

Experiment 5a. The effect of pH on the hydrolysis of starch by saliva

(a) Collect saliva as follows:
   (i) Rinse the mouth with water to remove food residues.
   (ii) Chew paraffin wax to promote the flow of saliva.
   (iii) Collect about 50 mm saliva in a clean test-tube.

(b) Draw up the table given below, in your notebook.

(c) Label five test-tubes 1-5.

(d) Using a graduated pipette, place 5 cm^3 1% starch solution in each tube.

(e) Add acid or alkali to each tube as indicated in the table below, rinsing the graduated pipette when changing from sodium carbonate to ethanoic (acetic) acid.

(f) Place several rows of drops of dilute iodine solution on a spotting tile. Half fill a beaker with water for rinsing the dropping pipette later in the experiment.

(g) *Note the time* and, using a graduated pipette, add 1 cm^3 saliva to each tube. Shake the tubes to mix the contents.

(h) Using a clean dropping pipette, remove a sample from each tube in turn and let one drop fall on to one of the iodine drops. Rinse the pipette.

(i) Continue sampling at intervals, taking samples most frequently from the tubes which are beginning to give a purple or reddish colour with the iodine. When a sample fails to give a blue or purple colour with iodine, *note the time* and stop taking samples from this tube.

(j) If after fifteen minutes some samples are still giving a blue colour with iodine, there is little point in continuing to test the mixture in these tubes.

(k) Rinse the dropping pipette and use it to remove a sample from tube 1. Touch the tip of the pipette on to a piece of pH test paper so that a *small* quantity of liquid runs on to it. Compare the colour produced on the paper with the standard chart supplied. Repeat this for each tube, rinsing the pipette between samples.

(l) With a fresh piece of pH test paper, find the pH in your mouth.

(m) Complete the table of results.

Tube	Starch solution and saliva plus:	pH	Time for blue colour to cease appearing
1	1 cm^3 sodium carbonate solution		
2	0.5 cm^3 sodium carbonate solution		
3	Nothing		
4	2 cm^3 ethanoic acid		
5	4 cm^3 ethanoic acid		

Experiment 5a. Discussion

1 At which pH was the hydrolysis of starch by saliva most rapid?

2 Is this the pH you would expect to be most effective? Explain.

3 If the conditions in the stomach are acid, what would happen to the reaction between saliva and the starch in a piece of bread when the bread was swallowed?

4 Suggest ways in which the sodium carbonate and acetic acid could influence the reaction between starch and saliva apart from merely altering the pH.

5 How could you extend the experiment to try and eliminate the possibility that the acid and alkali were affecting the reaction in some way other than through pH change?

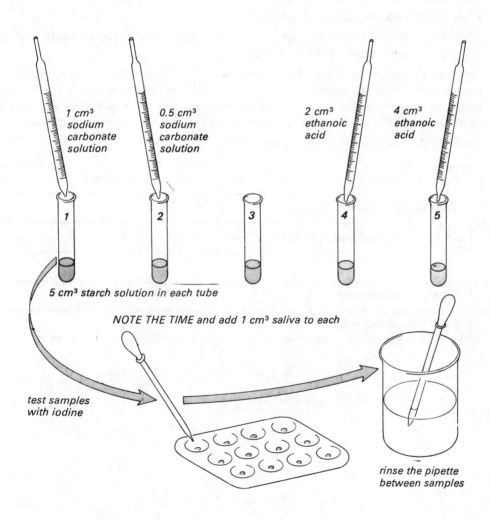

1 cm³ sodium carbonate solution

0.5 cm³ sodium carbonate solution

2 cm³ ethanoic acid

4 cm³ ethanoic acid

1   2   3   4   5

5 cm³ starch solution in each tube

NOTE THE TIME and add 1 cm³ saliva to each

test samples with iodine

rinse the pipette between samples

Experiment 5b. The effect of pH on the reaction between pepsin and egg-white

(a) Label five test-tubes 1-5.

(b) Using a graduated pipette place 5 cm³ of egg-white suspension in each tube.

(c) Using a graduated pipette, add acid or alkali to the tubes as indicated in the table below.

(d) Prepare a water bath in a 250 cm³ beaker or jar by mixing hot and cold water from the tap to give a temperature of about 40 °C. Have the beaker half full.

(e) Using a graduated pipette, add 1 cm³ 1% pepsin solution to each tube.

(f) Place all five tubes in the water bath.

(g) Copy the table below into your notebook.

(h) After five minutes, return the tubes from the water bath to the test-tube rack and compare the appearance of the contents.

(i) Compare the pH of each tube by taking a sample with a clean dropping pipette and touching the tip of the pipette on to a piece of pH test paper so that a *small* drop of liquid runs on to it. Compare the colour produced on the paper with the standard chart supplied. Repeat this for each tube, rinsing the pipette between samples.

(j) Record your results in the table in your book.

Tube	Egg-white suspension and pepsin plus :	pH	Appearance of contents after five minutes
1	2 cm³ sodium carbonate solution		
2	0.5 cm³ sodium carbonate solution		
3	Nothing		
4	1 cm³ hydrochloric acid		
5	2 cm³ hydrochloric acid		

Experiment 5b. Discussion

1 In the experiment, at what pH was the hydrolysis of egg-white most effective?

2 This pH is not necessarily the optimum (best) pH for the hydrolysis of egg-white by pepsin. Why not?

3 Can you determine from your results the pH least favourable to the reaction of pepsin on egg-white? Explain.

4 Suggest ways in which the sodium carbonate and hydrochloric acid could influence the reaction between pepsin and egg-white apart from merely altering the pH.

5 Do you think that the pH in your stomach corresponds approximately to the pH in the experiment which resulted in the most rapid hydrolysis of egg-white? If not, why not?

6 See if you can find out from a book what the pH in the human stomach really is.

7 Would you expect it to be the same in all people and at all times? Explain.

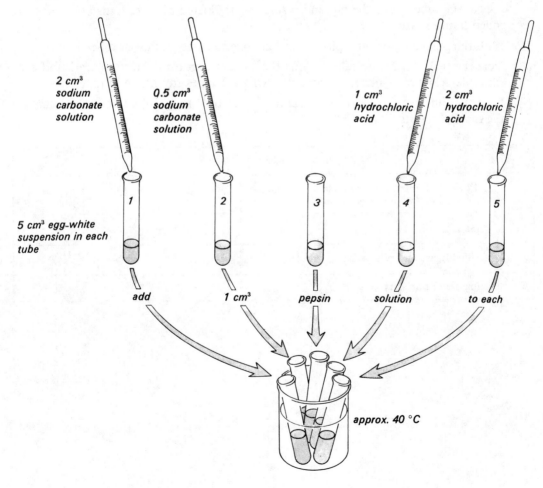

Experiment 6.   The action of lipase

There are three or four liquids to be added to each test-tube, in various combinations, so look at the table at the foot of this page to get an idea of the final contents of each tube.

(a)   Label three test-tubes 1-3.

(b)   Using a graduated pipette, place 5 cm³ milk in each tube.

(c)   Rinse the pipette and use it to place 7 cm³ dilute (M/10) sodium carbonate solution in each tube. This solution is to make the mixture alkaline.

(d)   Rinse the pipette and use it to place 1 cm³ 3% bile salts solution in tubes 2 and 3 only.

(e)   Use a dropping pipette to add phenolphthalein solution to each tube until the contents are bright pink. About six drops will be sufficient and equal quantities should be added to each tube.

  Phenolphthalein is a pH indicator. In alkaline solutions (above pH 10) it is pink; in 'acid' solutions (below pH 8.3) it is colourless.

(f)   In a spare test-tube, place about 15 mm of 5% lipase solution and, using a test-tube holder, heat the liquid over a small Bunsen flame until it boils for a few seconds.

  Cool the tube under the tap and, using the graduated pipette, transfer 1 cm³ of the boiled liquid to tube 2.

(g)   With the graduated pipette, place 1 cm³ unboiled lipase solution in tubes 1 and 3.

(h)   Note the time. Shake the tubes to mix the contents, return them to the rack and copy the table below into your notebook, observing the tubes from time to time.

(i)   Note the time required for the contents of each tube to go white and then complete the table of results.

	Action of lipase on milk	
Tube	*All three tubes contain milk, sodium carbonate and phenolphthalein plus :*	*Time taken to change from pink to white*
1	lipase only	
2	boiled lipase and bile salts	
3	lipase and bile salts	

Experiment 6. Discussion

1 What *food* substances are present in milk?

2 If phenolphthalein changes from pink to colourless, what kind of chemical change must have taken place in the tube?

3 Recall (or look up) the final products of digestion of the principal classes of food and write down which of these products could be formed by the digestion of milk.

4 Which of the final products of digestion of milk could be responsible for the change of conditions in the test-tube?

5 Which part of the experiment suggests that lipase acts as an enzyme?

6 What chemical change could the lipase be producing which would account for the colour change in the test-tubes?

7 Which part of the experiment indicates that bile salts do not contain an enzyme which affects milk (at least in the way being investigated here)? Explain your reasoning.

8 From the *results,* assuming that lipase is an enzyme, what part do the bile salts appear to be playing in the reaction (in general terms)?

9 Do the results tell you whether lipase is acting on the fat or the protein in milk? Explain.

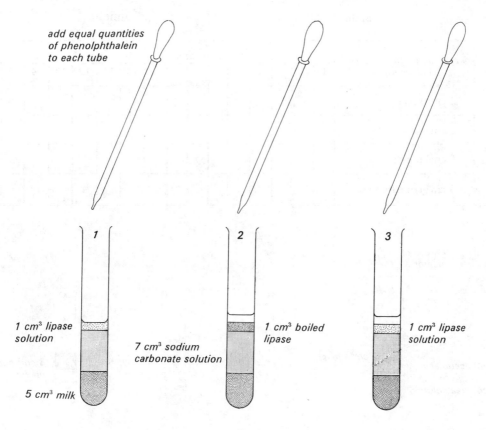

add equal quantities of phenolphthalein to each tube

1

2

3

1 cm³ lipase solution

7 cm³ sodium carbonate solution

1 cm³ boiled lipase

1 cm³ lipase solution

5 cm³ milk

Experiment 7. The effect of the concentration of enzyme on the rate of reaction

(a) Copy the table given below into your notebook.

(b) Label four test-tubes 1-4.

(c) Using a graduated pipette, add 5 cm³ 1% urea solution to each tube.

(d) Rinse the pipette and use it to add 2 cm³ dilute acetic acid to each tube.

(e) Add ten drops of BDH universal indicator to each tube, using the dropper incorporated in the bottle itself or a dropping pipette.

(f) Rinse the graduated pipette and use it to put 3 cm³ *boiled* urease solution into tube 4.

(g) Now place the following volumes of urease (unboiled) in tubes 1-3:
    (1) 2 cm³,   (2) 3 cm³,   (3) 5 cm³.

(h) At intervals of 30 seconds shake the tubes, observe the colour of the indicator in each tube and record the corresponding pH in your table (see the key below).

BDH Universal Indicator	pH	
PINK	4	(acid)
ORANGE	5	
YELLOW	6	
GREEN	7	
BLUE-GREEN	8	(alkaline)

Tube	5 cm³ urea solution, 2 cm³ acetic acid plus:	1	2	3	4	5	6	7	8	9	10	11	12	13	14	15	16
		*Time intervals (30 second)*															
1	2 cm³ urease																
2	3 cm³ urease																
3	5 cm³ urease																
4	3 cm³ boiled urease																

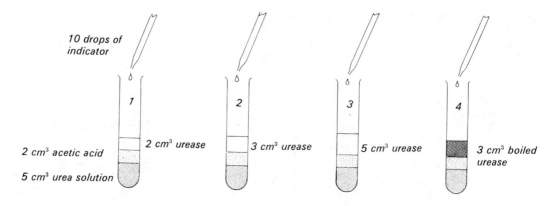

10 drops of indicator

2 cm³ acetic acid
5 cm³ urea solution

1 — 2 cm³ urease
2 — 3 cm³ urease
3 — 5 cm³ urease
4 — 3 cm³ boiled urease

Experiment 7. Discussion

1 What change in pH took place in tubes 1-3? Is the change from acid to alkali or the reverse?

2 What evidence is there to suggest that the change was brought about by an enzyme?

3 What was the effect of increasing the amount of enzyme on the rate of reaction? Would you expect this effect to continue indefinitely with an increasing enzyme concentration or would you expect a limit to be reached? Explain your answer.

4 The acetic acid was added to make the conditions artificially acid to start with. What substance could urease be producing from urea which could make the conditions alkaline?

$$\text{The formula for urea is } (NH_2)_2CO \text{ or } \begin{matrix} NH_2 \\ \\ NH_2 \end{matrix} \!\!\!\!> CO$$

5 What part could the enzyme urease be playing in the metabolic process of plants?

6 In general, of what value to a living organism could be the effect of an increased rate of reaction with increasing enzyme concentration?

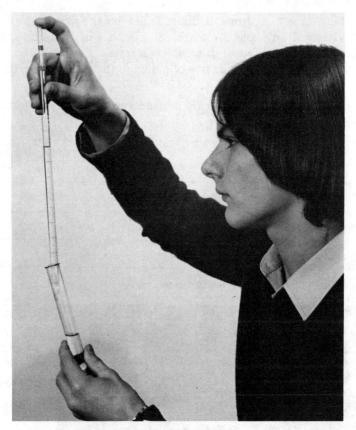

Add 5 cm³ urea solution to each tube

Experiment 8.   The action of starch phosphorylase

(a) Cut a piece of potato about 20 mm square and further cut it into small pieces. Place these in a mortar with a little clean sand and grind them finely with a pestle.

(b) Using a graduated pipette, add 10 cm^3 distilled water to the mixture and continue grinding for a few seconds.

(c) Filter the extract through a fluted filter paper, into a clean test-tube.

(d) While you are waiting for the filtration, use a dropping pipette to place eight drops of a 5% solution of glucose phosphate on a spotting tile, in two rows of four drops.

(e) Rinse the dropping pipette and use it to draw up some of the filtered potato extract. Place one drop in a spare cavity in the tile and test it with a drop of iodine. If it goes blue, the extract will have to be filtered again.

(f) If the sample of filtrate does not give a blue colour when tested with iodine, place one drop of the extract on each drop of glucose phosphate in the *top* row.

(g) Using a test-tube holder, boil the remainder of the potato extract in the test-tube, over a small Bunsen flame; rinse the dropping pipette and use it to place a drop of the boiled extract on each drop of glucose phosphate in the *bottom* row. Note the time.

(h) Copy the table below into your notebook.

(i) After 5 minutes place four drops of dilute iodine on the *first* drop of the mixture in each row and record any colour change. At intervals of 5 minutes, repeat this test with the second, third and fourth drops in both rows.

   From time to time add a drop or two of iodine to the samples already tested because the colours tend to fade.

(j) When the experiment is complete, use a spare cavity in the tile to test a sample of glucose phosphate with iodine.

	Drops	Colour on adding iodine at 5 min intervals			
		5 *min*	10 *min*	15 *min*	20 *min*
Top row	Potato extract and glucose phosphate				
Bottom row	Boiled potato extract and glucose phosphate				
Colour on adding iodine to glucose phosphate alone:					
Colour on adding iodine to filtered potato extract alone:					

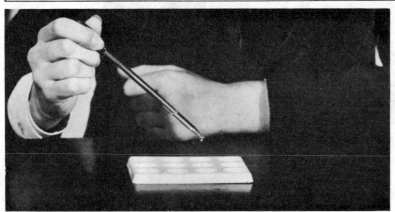

Place 1 drop of glucose phosphate in each of 8 cavities

Experiment 8.   Discussion

1 Does the filtered potato extract contain any starch?

2 Does the glucose phosphate contain any starch?

3 How do you interpret the colour changes that occur when iodine is added to the mixture of glucose phosphate and potato extract?

4 Consider two possibilities:
   (i) something in potato extract has converted glucose phosphate to starch.
   (ii) the glucose phosphate has converted something in potato extract to starch.
   (a) In view of the results with the second row of drops, which of the two possibilities is the more likely? Say why.
   (b) Suggest a further control experiment that might help to support one of the two alternatives.

5 Are your results consistent with the hypothesis that the potato extract contains an enzyme which converts glucose phosphate to starch?

6 Give a simple chemical explanation of how this change could come about assuming that the phosphate merely makes the glucose more reactive. (Consult the structural formulae on p. 39.)

7 What part could such an enzyme play in the development of a potato tuber?

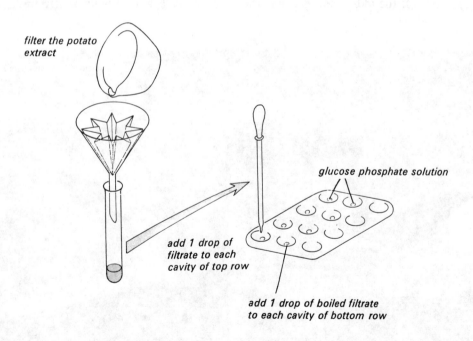

filter the potato extract

glucose phosphate solution

add 1 drop of filtrate to each cavity of top row

add 1 drop of boiled filtrate to each cavity of bottom row

Experiment 9. Starch-digesting enzymes in maize fruits

You are provided with the following:
    (i) A clean Petri dish.
    (ii) A test-tube containing a dilute starch solution and melted agar.
    (iii) Two maize fruits which have been soaked for at least twenty-four hours. One has been killed by boiling; both have been soaked in sodium hypochlorite solution to destroy any fungi on their surface.

(a) Take a test-tube of starch-agar from the water bath and, without delay, pour it into the lower half of the Petri dish as shown in Fig. 1. Replace the lid at once and *allow the agar to cool and set* (at least five minutes).

(b) While the agar is cooling, cut the two maize fruits in half longitudinally as shown in Fig. 2.

(c) When the agar has set, use forceps to place the half fruits carefully, cut surface downwards on the agar, arranging them as shown in Fig. 3 so that the boiled and unboiled fruits can be identified later.

(d) Replace the lid of the Petri dish, mark your initials on it with a spirit marker or glass-writing crayon, and leave for one to seven days.

(e) After the appropriate time interval, remove the maize fruits and pour dilute iodine solution into the dish to cover the agar surface for one or two minutes.

(f) Pour off the iodine and examine the agar surface against a light background.

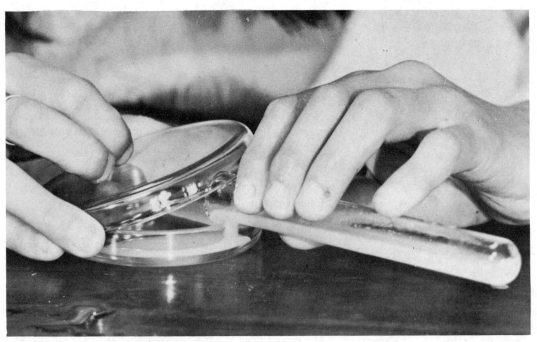

Fig. 1. Raise the lid as little as possible

Experiment 9. Discussion

1 What is the significance of a blue colour when iodine solution comes in contact with the agar?

2 In which parts of the agar did a blue colour fail to appear?

3 Since the starch-agar covered the base of the Petri dish and, presumably, the starch was evenly distributed throughout the agar, what can you assume about the areas which fail to go blue?

4 Offer an explanation which connects the colourless area of agar and the unboiled maize fruits.

5 What difference is there in the agar beneath the boiled and unboiled maize fruits after the iodine has been added?

6 How do you account for this difference, in terms of your answer to 4?

7 Are your results consistent with the hypothesis that living maize fruits produce an enzyme which digests starch?

8 What further information about the contents of the agar would help you to decide?

9 If this theory is acceptable, what part would such enzymes play in the germination of a maize fruit?

10 Do the results rule out the possibility that the living maize fruits have, in some way, simply absorbed the starch from the agar?

11 Suggest an experiment which would help to decide between hypotheses 7 and 10.

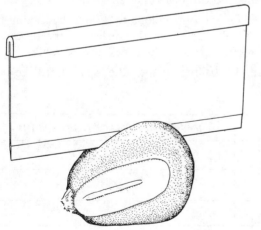

Fig. 2. Cut the maize fruit longitudinally

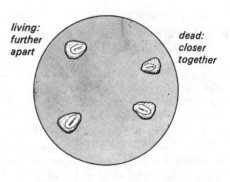

Fig. 3. Position of grains on the agar

Experiment 10. Dehydrogenase in yeast

During respiration, hydrogen atoms are removed from glucose molecules by enzymes called dehydrogenases and passed to various chemicals called hydrogen acceptors. As the hydrogen atoms pass from one hydrogen acceptor to another, energy is made available for chemical reactions in the cell. In this way, substances such as glucose provide energy for vital reactions in living organisms.

In this experiment, a dye called methylene blue acts as an artificial hydrogen acceptor. When this dye is reduced by accepting hydrogen atoms it goes colourless.

(a) Place about 30 mm of yeast suspension in a test-tube and, using a test-tube holder, heat this suspension over a small Bunsen flame until the liquid boils for about half a minute. Then cool the tube under the tap.

(b) Label three test-tubes 1-3.

(c) Using a graduated pipette, place 2 cm³ of the boiled yeast suspension in tube 1.

(d) Using the graduated pipette, draw up 4 cm³ unboiled yeast suspension and place 2 cm³ in tube 2 and 2 cm³ in tube 3.

(e) Rinse the pipette and use it to place 2 cm³ distilled water in tubes 1 and 2.

(f) With the pipette, place 2 cm³ 1% glucose solution in tube 3.

(g) Prepare a water bath by mixing hot and cold water from the tap to obtain a temperature between 35 and 45 °C. Place all three tubes in this water bath for 30 minutes. Rinse the pipette.

(h) Copy the table given below into your notebook.

(i) After five minutes draw up 6 cm³ methylene blue solution in the pipette and place 2 cm³ in each tube. Shake all three tubes thoroughly for about 20 seconds and return them to the water bath, noting the time as you do so. Do not shake the tubes again.

(j) Watch the tubes to see how long it takes for the blue colour to disappear, leaving the creamy colour of the yeast. A thin film of blue colour at the surface of the tube may be ignored but the tubes should not be moved. Record the times in your table.

(k) The experiment may be repeated by simply shaking all the tubes again until the blue colour returns.

Tube	Contents	Time for methylene blue to go colourless
1	Boiled yeast	
2	Unboiled yeast	
3	Unboiled yeast + 1% glucose	

Experiment 10.   Discussion

1 Why was distilled water added to tubes 1 and 2?

2 What causes the methylene blue solution to go colourless (according to the introduction on p. 64)?

3 How do you explain the results with tube 1?

4 In which of tubes 2 and 3 was the methylene blue decolourized more rapidly? How can this result be explained?

5 If the hydrogen atoms for the reduction of methylene blue come from glucose, why should the methylene blue in tube 2 become decolourized at all?

6 What do you think would be the effect of increasing the glucose concentration in tube 3? Explain your answer.

7 How could you extend the experiment to see if *enzymes* in yeast are capable of reducing methylene blue?

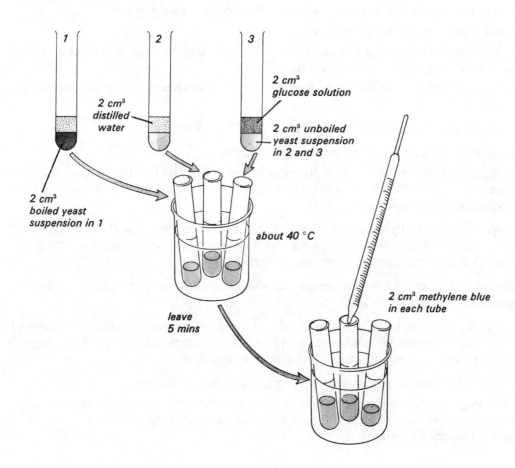

2 cm³ glucose solution

2 cm³ distilled water

2 cm³ unboiled yeast suspension in 2 and 3

2 cm³ boiled yeast suspension in 1

about 40 °C

leave 5 mins

2 cm³ methylene blue in each tube

Experiment 11. Catalase

Catalase is an enzyme which occurs in the cells of many living organisms. Certain of the energy-releasing reactions in the cell produce hydrogen peroxide as an end-product. This compound, which is toxic to the cell, is split to water and oxygen by the action of catalase.

$$2H_2O_2 = 2H_2O + O_2$$

The investigation below is a fairly critical examination of plant and animal tissues to see if they contain catalase.

    (a) Label three test-tubes 1-3.

    (b) Pour about 20 mm hydrogen peroxide into each tube.

    (c) Cut the liver into 3 pieces.

    (d) To tube 1 add a small piece of liver, and to tube 2 add a pinch of dried yeast.

    (e) Insert a glowing splint into tubes 1 and 2, bringing it close to the liquid surface or into the upper part of the froth.

1 Describe what you saw happening and the effect on the glowing splint.

2 How do you interpret these observations?

3 Is there any evidence from this experiment so far, to indicate whether the gas is coming from the hydrogen peroxide or from the solid?

4 Is there any evidence at this stage that an enzyme is involved in the production of gas in this reaction?

    (f) In tube 3 place a few granules of charcoal and observe the reaction.

5 Could charcoal be an enzyme? Explain your answer.

6 Assuming **(i)** that the gas in **(f)** is the same as before and **(ii)** that the charcoal is almost pure carbon, does the result with charcoal help you to decide on the source of the gas in this and the previous experiments?

    (g) Suppose the hypothesis is advanced that there is an *enzyme* in the liver and yeast, which decomposes hydrogen peroxide to oxygen and water; design and carry out a control experiment to test this hypothesis.

7 Record **(i)** the experiment, **(ii)** the reasons which led you to conduct it, **(iii)** the observed results and **(iv)** your conclusions.

    (h) Wash out the test-tubes. Design and carry out an experiment to see if the supposed enzyme in the plant and animal material can be extracted and still retain its properties. The experiment should include a control.

8 Describe briefly your procedure, your results and your conclusions.

9 Assuming that liver and yeast each contain an enzyme which splits hydrogen peroxide, is there any evidence to show that it is the same enzyme? What would have to be done to find this out for certain?

3 Soil

Introduction

The experiments in this section attempt to give some insight into the composition and properties of soil. Experiments 1, 2, 3, 12 and 13 are concerned with measuring the water, air, organic and mineral content of the soil. Although the figures derived from these experiments will give a general idea of the composition of soil, they will be more informative if strongly contrasted soils are sampled and the results used as a basis for comparison. Experiments 4, 5, 6, 7 and 9 do involve a comparison of certain properties of natural soils or artificial mixtures, while experiments 10, 11 and 14 investigate the presence and activities of living organisms in the soil. Nine of the experiments can be completed in a double period or less but experiments 1, 6, 10, 11 and 14 need several days to give satisfactory results.

Contents

Experiment 1. To find the percentage of water in a sample of soil

(a) You are provided with some freshly collected, sieved soil.

(b) Weigh out exactly 100 g and place it in an evaporating basin or other container.

(c) EITHER write your name on the container, using a spirit marker, OR write your name on a piece of paper and place it in the container with the soil.

(d) Place the container with the soil in an oven kept at 105 °C and leave it there for 24 hours.

(e) If soil from a different site is available, repeat the experiment from (a).

(f) After 24 hours, using a pair of tongs if the oven is still hot, remove the container and soil from the oven and allow to cool. Tip the soil into the scale pan of the balance and weigh it. (Do not throw the dry soil away; it will be used for experiment 2.)

(g) Record the weight of dry soil in your notebook and calculate the percentage of water in the sample as follows:

First weight of soil = 100 g
Second weight of soil = ... g
Loss in weight = 100 — ...... = ... g

This loss in weight has resulted from the evaporation of water from the soil,

therefore, 100 g soil contained ...... g water
and the percentage of water in this sample is ...... %

Experiment 1. Discussion

1 The soil could have been dried more quickly by heating it over a Bunsen burner to a temperature *far* above 105 °C.

   Why would such a technique be unsatisfactory, i.e. what errors would it involve?

2 It has been assumed that after twenty-four hours in an oven at 105 °C, all the water had been evaporated from the soil. If some water still remained, how would it affect the results, i.e. would the calculated percentage of water be too high or too low?

3 What would you have to do to be quite sure that all the water had been driven off by heating?

4 Suppose that one of your samples contained a large proportion of vegetable matter, e.g. roots or dead leaves, how would this affect your results?

5 The percentage of water in the soil will vary according to the extent to which it has been wetted by recent rainfall or dried by sunshine and plant roots. In view of this variability what is the point, if any, of attempting to measure the percentage of water?

   If you think there is some value in the experiment, what uses could a scientist make of the results?

   (Imagine an irrigation programme in which the farmer has to pay for the water he uses.)

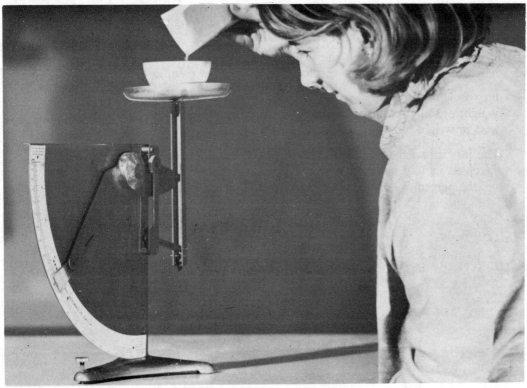

Weigh out 100 g soil

Experiment 2. To find the percentage of organic matter in a sample of soil

The organic matter in the soil consists, usually, of dead and decaying plant and animal remains.

The sample of soil *must be dry*. It is usually convenient to use the dried soil from experiment 1, p. 70.

(a) You are provided with some dried soil.

(b) Use a balance to weigh out exactly 50 g of the dried soil and place it in a metal container.

(c) Put a protective mat on the bench and then place the tray of soil on a tripod over the mat. The mat is used to protect the bench from the heat 'reflected' from the bottom of the tray.

(d) Heat the tray from below as strongly as possible with a Bunsen burner for fifteen minutes. The tray and soil may become red-hot.

(e) During the heating, move the soil about with a metal spatula so that it all comes in contact with the hottest part of the tray for a time. Notice the production of smoke and the changes in the colour of the soil.

(f) If you are going to repeat the experiment with another soil sample, you should be weighing it out while heating the first sample, to save time.

(g) After fifteen minutes heating, allow the tray to cool, tip the soil on to the scale-pan of the balance and record the new weight.

(Question 5 asks you what is the smallest unit of weight marked on the scale of the balance. Make a note of this now.)

(h) If you are going to repeat the experiment with another soil sample, start heating it now while you calculate the results of the first experiment as below:

First weight of soil = 50 g
Weight of soil after heating strongly = ... g
Loss in weight = 50 — ...... = ... g

The loss in weight represents the weight of organic matter which has been burned, therefore, in 50 g soil there was ...... g organic matter
and the percentage organic matter in this soil sample is ...... × 2 = ... %

Experiment 2. Discussion

1 If the soil sample had not been completely dry to start with, how would this have affected your results?

2 What is the reason for heating the soil so strongly and why should this result in a loss of weight?

3 What were the colour changes which you observed during the heating and subsequent cooling of the soil? (Ignore the 'red-hot' appearance during heating.) Suggest an explanation for these changes.

4 It is possible that fifteen minutes heating is not sufficient to achieve the maximum loss of weight in the soil. What would you have to do to be quite sure that the maximum weight loss had been achieved?

5 What is the smallest unit of weight marked on your balance? If you mis-read the scale by one of these units, how great would be the error in your results? If you had worked with only 10 g soil, would the same mistake in reading have produced the same error?

6 If you were to analyse two or more samples of soil for their organic content, what would you regard as a significant difference between the results, i.e. a difference greater than would be expected from a simple experimental error?

7 Suggest an explanation for the difference, if significant.

Move the soil about with a metal spatula

Experiment 3. To find the percentage of air in a sample of soil

You are provided with a large beaker for water and two cans of equal volume. One of the cans is perforated at the base.

1 Collecting the sample (see Fig. 2, p. 76)

Use a spade or trowel to remove the top centimetre of turf, plant material or debris from a patch of moist soil. Drive the perforated can, open end downwards, into the soil by pushing or stamping gently on it. As far as possible avoid compressing or otherwise disturbing the soil.

When the base of the can is level with the soil and the soil is visible against the perforated base, dig the can out carefully with the spade or trowel, without dislodging the soil from the can.

Cut the soil level with the top of the can.

2 Measuring the volume of air (see Fig. 1, p. 75)

(a) Put the empty, non-perforated can into the beaker and fill the beaker up to the mark with cold water.

(b) Remove the can upright and full of water. The level in the beaker will drop by an amount equivalent to the volume of the can. (There is no need to measure this drop in level.)

(c) Put the can of soil into the beaker of water. The water will at first return to its original level, because the volume of soil is equal to the volume of water just removed. As the air escapes from the soil, however, the volume will drop. The fall in level will be equivalent to the volume of air which escapes from the soil.

(d) With a spatula or stick, loosen the soil in the can to allow all the air to escape. The can may be removed from the water to do this, provided that all the soil and finally the can itself are returned to the water.

(e) Fill a measuring cylinder with water to its upper graduation, e.g. 100 cm³. Pour this water into the beaker until the water returns to its original level. Note the volume of water left in the measuring cylinder. The difference between the first and second volume will give you the volume of water added to the container.

(f) The volume of water added to the container is equal to the volume of air which escaped from the soil.

(g) Use the measuring cylinder to find the volume of the non-perforated can. This will give you the volume of soil used.

(h) Write the results in your notebook and from them calculate the percentage of air in the soil sample.

Results

Volume of soil (i.e. the volume of the can) = ... cm³
Volume of air (i.e. 1st reading on measuring cylinder — 2nd reading) = ... cm³

$$\text{Percentage of air in soil sample} = \frac{\text{volume of air} \times 100}{\text{volume of soil}}$$

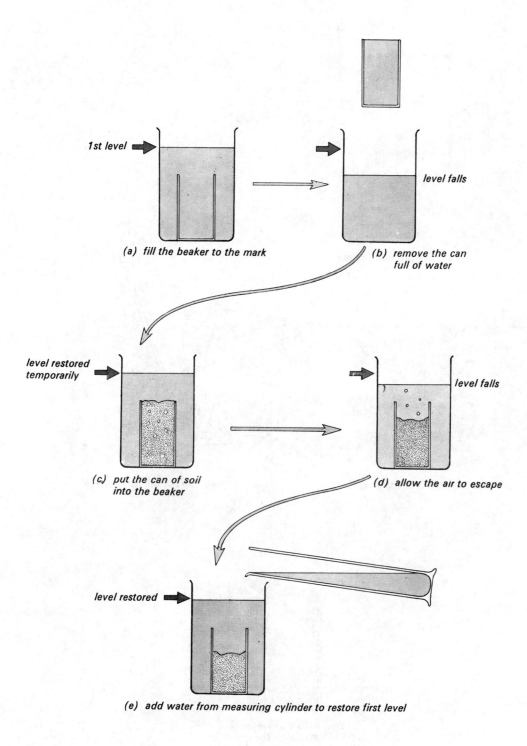

1st level ➡

level falls

(a) fill the beaker to the mark

(b) remove the can full of water

level restored temporarily ➡

level falls

(c) put the can of soil into the beaker

(d) allow the air to escape

level restored ➡

(e) add water from measuring cylinder to restore first level

Fig. 1. Measuring the volume of air in a soil sample

75

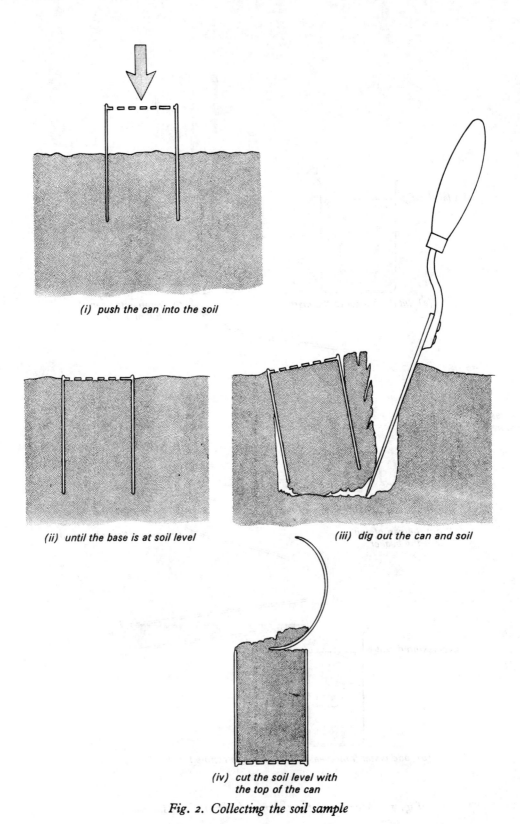

(i) push the can into the soil

(ii) until the base is at soil level

(iii) dig out the can and soil

(iv) cut the soil level with
the top of the can

Fig. 2. Collecting the soil sample

Experiment 3. Discussion

1 When the can of soil is lowered into the beaker of water, why does the water return to its first level?

2 Why doesn't the water remain at this level?

3 How did you find what the fall in level was, in terms of volume?

4 What effect would it have on your results if
   (a) the soil was compressed when the sample was taken,
   (b) the sample included a large stone,
   (c) the sample was taken by shovelling soil into the can with the aid of a trowel?

5 How would you expect the results to vary after (a) heavy rainfall, (b) prolonged drought?

6 What is the volume indicated by the smallest division on the measuring cylinder? Suppose that, in reading the volume, you misjudged the level by one of these divisions, would this make your results very inaccurate?

7 Apart from the measuring cylinder, what other sources of inaccuracy are there in the experiment?

8 Why cannot the percentage of air found by this experiment be compared with the percentages of water and organic matter found in experiments 1 and 2?

Pour water from a measuring cylinder to restore the original level in the beaker

Experiment 4.   A comparison of the permeability to water of different samples of subsoil

(a) Copy the table given below into your notebook.

(b) You are provided with three funnels each containing equal weights of different mixtures of sand and clay at *field capacity*, i.e. containing as much water as can be held against gravity.

(c) Support one of the funnels in a clamp or tripod so that its stem leads into a small measuring cylinder.

READ INSTRUCTIONS (d) to (f) before continuing.

(d) NOTE the time and pour water from a beaker into the funnel until the water reaches the line marked on the funnel.

(e) As the water passes through the mixture, keep adding water to the funnel to maintain the level.

(f) After one minute (or any whole number of minutes) note the volume of water that has run into the measuring cylinder and the time taken for this volume to collect. Record the time and volume in your table of results. (Choose a number of minutes that gives an easily readable volume of water in the measuring cylinder.)

(g) Empty the measuring cylinder and repeat the experiment using the second and third samples.

(h) Calculate the rate at which water passes through each sample, in cm^3 per min.

Composition of sample		Time for water to run through	Volume of water collected	Volume of water per min passing through mixture
% Clay	% Sand			
0	100			
5	95			
15	85			

Experiment 4.   Discussion

1 Why was the water in the funnel kept at the same level throughout the experiment?

2 Which sample was most permeable to water?

3 What is the effect of an increasing proportion of clay on the permeability of the soil mixture to water?

4 In what ways do you think this experiment is unrepresentative of what actually happens when water drains through topsoil?

5 Suggest one possible advantage and one possible disadvantage of a soil which was relatively impermeable to water.

6 What results do you think you might have obtained if the sample of soil in the funnel contained pure clay?

7 What difference would it make to the results if (a) one funnel was wider than the others but had the same depth of soil in it, (b) one funnel had a deeper layer of soil than the others?

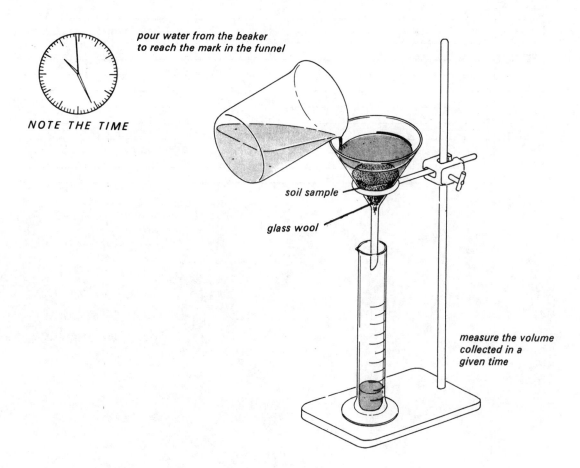

pour water from the beaker to reach the mark in the funnel

NOTE THE TIME

soil sample

glass wool

measure the volume collected in a given time

79

Experiment 5. The permeability of topsoil to water

You are provided with a number of funnels containing equal depths of soil of different kinds.

(a) Copy the table given below into your notebook and fill in the particle size of each sample as you test it.

(b) Place one of the funnels in a ring or clamp so that the stem is directed into a measuring cylinder.

READ INSTRUCTIONS (c) and (d) before continuing.

(c) Note the time and pour water from a beaker into the funnel up to the top mark. As the water drains through the soil add more water to keep the level constant.

(d) After exactly 15, 30 or 60 seconds, according to the rate at which the water is collecting, remove the funnel from the measuring cylinder and note the volume of water which has passed through the soil. If necessary, multiply this volume by 2 or 4 to obtain the number of cm^3 passing per minute and record this figure in the table.

(e) Empty the measuring cylinder and repeat the experiment with the other soil samples.

Sample	Range of particle size	Volume of water passing per minute
Sand		
Clay		
Loam		

Experiment 5. Discussion

1 Choose a suitable scale (e.g. 1 mm = 2 cm³) and draw a block graph of the volume of water passing through each sample in one minute. Use colours or shading to identify the blocks referring to sand, clay and loam.

2 Which sample was (i) most permeable, (ii) least permeable, to water?

3 The results of experiment 4 probably suggested that the more clay present in an artificial soil mixture, the less permeable it was to water. From your results with experiment 5 can you say that a clay soil will be less permeable to water than a sandy soil? Justify your answer.

4 From your results, discuss which you think is more important in determining permeability, the soil's composition or the size of its constituent particles (crumbs).

5 In what ways is this experiment unrepresentative of real conditions in the soil?

6 Why was it necessary during the experiment to keep the water in the funnel at a constant level?

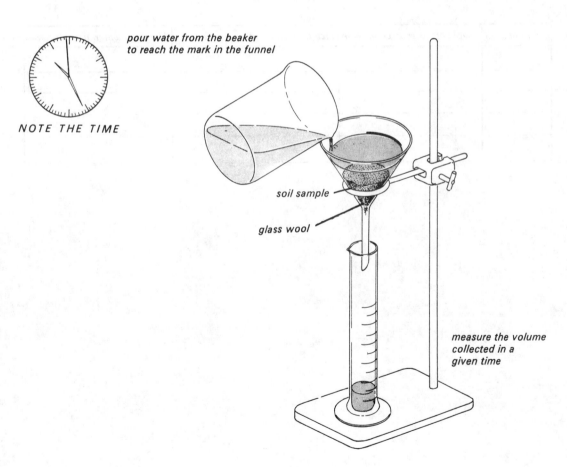

NOTE THE TIME

pour water from the beaker
to reach the mark in the funnel

soil sample

glass wool

measure the volume
collected in a
given time

Experiment 6. Capillary attraction and particle size

You are provided with two or three long glass tubes plugged at one end with glass wool, and two or three samples of sieved sand, graded according to particle size.

(a) Label the top of each tube with the grade of sand it is to contain.

(b) Use a dry plastic funnel to fill each tube to the top with the appropriate grade of sand. Tap the tubes lightly on the bench to compact the sand.

(c) Fill a beaker or jam jar with water to a depth of about 50 mm and clamp the tubes vertically with the plugged end in the water.

(d) Copy the table given below into your notebook.

(e) By the end of the lesson, the water will have travelled some distance up the columns of sand. Since the wet sand looks darker, the level reached by the water can be measured. It is convenient and consistent to put the ruler in the water and let it rest on the bottom of the beaker when measuring the height of water in the sand columns. Record the results in your table.

(f) If possible, make further measurements one or two days later.

(g) Measure and record the heights after one week.

Grade of sand	Height of water after one hour	Height of water after ...... day(s)	Height of water after one week

Experiment 6. Discussion

1 What relationship, if any, is there between the size of sand particle and the height of the capillary rise?

2 Were the results after one hour in the same order as the results after one day and one week?

3 In what ways is capillary attraction likely to affect the movements and distribution of water in a soil?

4 From the experiment, it is clear that capillary attraction represents a force which can pull water upwards against gravity. When roots extract water from between soil particles what forces must they overcome?

5 From which type of soil (coarse or fine particles) should plant roots extract water more easily? Explain.

6 What effect would you expect capillary attraction to have on the water-retaining properties of a soil?

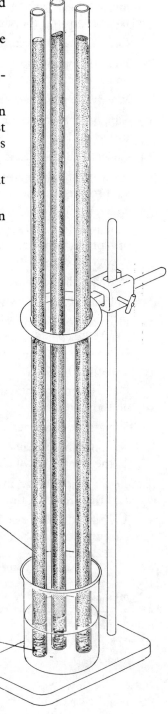

graded sand

glass wool

Experiment 7. Retention of water

You are provided with two plastic containers, perforated at the base and containing a disc of glass fibre to stop the samples falling through; 100 cm^3 sand and 100 cm^3 dry peat.

(a) Copy the table given below into your notebook.

(b) Pour the sand into one of the plastic containers; it should reach the 100 cm^3 mark on the side.

(c) Weigh the container with the sand and record the weight in your table.

(d) Support the plastic container in a ring or clamp and place a beaker or jam jar underneath it.

(e) Collect 100 cm^3 water in a measuring cylinder and pour it into the sand.

(f) Allow to stand for five minutes for the water to drain through. While you are waiting, start to prepare the sample of peat as described in (i).

(g) After five minutes, water will have stopped dripping from the sand. Remove the container from its support without shaking or compressing it and stand it on a piece of blotting paper on the bench for about two minutes, moving it to a dry patch every twenty seconds or so.

(h) Reweigh the container with the sand and record its weight in your table.

(i) Fill the measuring cylinder with 100 cm^3 water.

(j) Place the peat in a tray and add 30 cm^3 water from the measuring cylinder. Work the peat with your fingers to distribute the water and make the material uniformly moist. (If this is not done, the dry peat will not absorb water when it is poured into the container.)

(k) Place the moist peat in the plastic container and compress it to the 100 cm^3 mark on the side.

(l) Weigh the container with the peat and record its weight.

(m) Support the container in the ring or clamp, place a beaker or jar under it and pour in the remaining 70 cm^3 water from the measuring cylinder.

(n) Leave to drain for five minutes; place on blotting paper as before for two minutes and then reweigh the container.

(o) Calculate the increase in weight of each sample and hence the weight of water retained, not forgetting that the peat starts with 30 cm^3 water. (30 cm^3 water weighs 30 g.)

	1st weight (dry sample)	2nd weight (wet sample)	Increase in weight (water retained)
100 cm^3 sand			
100 cm^3 peat			30+

Experiment 7. Discussion

1 Which sample seemed to retain the more water? How much more did it retain?

2 What would you have to do to be sure that this difference was not just an experimental variation as might occur between two identical samples of sand subjected to the experimental procedure?

3 Would you expect a soil which was very permeable to water also to have a high level of water retention? Explain.

4 Do you think that a high level of water retention in a soil is likely to be beneficial or harmful to plants growing in the soil? Suggest reasons.

5 Organic matter is a source of mineral nutrients in the soil. What other advantages of organic matter are suggested by the results of this experiment?

6 In what ways does this experiment differ from experiment 1 which tries to find the percentage of water in a soil sample?

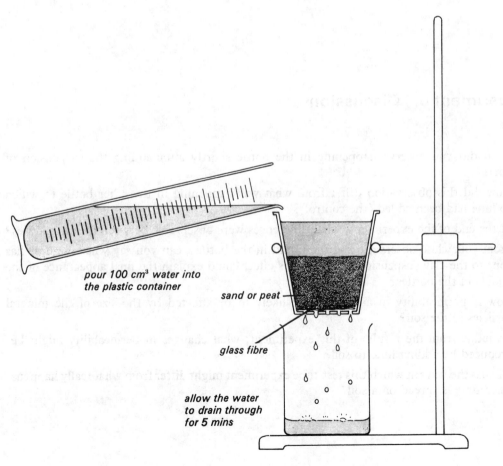

pour 100 cm³ water into
the plastic container

sand or peat

glass fibre

allow the water
to drain through
for 5 mins

Experiment 8.   The effect of lime on clay

(a) Place 3 g powdered clay on a piece of paper; fold the paper and use it to tip the clay into a medicine bottle. Repeat this with the second bottle.

(b) Fill both bottles with water to within a centimetre or two from the top, replace the stopper and shake the bottles vigorously to disperse the clay in the water.

(c) Stand the bottles on the bench.

(d) In a test-tube place about 2 mm slaked lime (calcium hydroxide); add about 50 mm water and shake the tube to make a suspension of lime.

(e) Pour this suspension into ONE of the bottles; replace the stopper and shake both bottles again.

(f) Place the bottles on the bench and look closely at the liquid in each bottle, watching for any differences **(i)** in the appearance of the suspended particles and **(ii)** in their movements.

(g) If possible, leave the two bottles undisturbed until the end of the lesson and note any difference in their appearance after this period of time.

Experiment 8.   Discussion

1 What did you observe happening in the bottle shortly after adding the suspension of lime?

2 How did this observation differ from what was happening in the other bottle to which no lime had been added (the control)?

3 At the end of the experiment, what differences were observable between the two bottles?

4 From your close observation of the liquid in the bottles, can you say what the lime has done to the clay suspension and how this effect could explain the final appearance of the liquids in the bottles?

5 How is permeability to air and water likely to be affected by the size of the mineral particles in the soil?

6 To judge from the results of this experiment, what changes in permeability might be produced by adding lime to soil?

7 Discuss the ways in which this test-tube experiment might differ from what really happens when lime is spread on a soil.

Experiment 9. The pH test on soil

(a) Place about 10 mm soil in the bottom of a test-tube.

(b) Add about 10 mm barium sulphate powder. This is to flocculate and precipitate the clay.

(c) Add 10 cm³ distilled water, using a graduated pipette.

(d) Use the pipette to add 2 cm³ soil indicator to the test-tube.

(e) Close the mouth of the test-tube with your thumb and shake the tube vigorously up and down to mix the contents.

(f) Return the tube to the rack and allow the contents to settle for a minute. A clear area will appear, whose colour gives an indication of the pH.

(g) Hold the clear coloured area against the chart of pH colours and decide which of the colours corresponds most nearly to the colour in the tube.

(h) Repeat the experiment with soil from a different source.

Source of soil	pH

Experiment 9. Discussion

1 What assumption must be made about the barium sulphate in this experiment?

2 Why should slaked lime not be used to precipitate the clay as in experiment 8?

3 Why is distilled water used in preference to tap water?

4 Assuming that variations in the volumes of soil used and the volume of water in which they are shaken do not greatly affect the measurement of pH, what factors limit the accuracy of this experiment?

5 By how much would the pH of soils have to differ, as judged from this test, for you to say confidently that the difference was not merely experimental error?

6 How would you have to modify this experiment to be able to give a gardener a value for the pH of soil in his garden?

7 How do you suppose a botanist would be able to recognize an acid soil or an alkaline soil without doing a pH test on it?

Experiment 10. Micro-organisms in the soil

(a) You are provided with two sterile Petri dishes and two test-tubes in each of which are 10 cm³ melted agar jelly containing nutrients on which micro-organisms can feed. Light the Bunsen burner and adjust it to give a non-luminous flame.

(b) Remove the foil or cotton-wool plug from one test-tube, pass the mouth of the tube two or three times through the Bunsen flame and pour the liquid jelly into the bottom half of a Petri dish while raising the lid carefully as shown in Fig. 1. Replace the lid and repeat the procedure with the second dish and tube of liquefied agar.

The agar must now be left to set solid (about five min).

(c) Use a spirit marker to write 'Experiment' on the lid of one dish and 'Control' on the other. Add your initials and the date to each dish.

(d) DO NOT CONTINUE WITH THE EXPERIMENT UNTIL THE AGAR HAS SET. Check for this by tilting each dish gently to see if the agar runs.

(e) Sterilize the transfer loop by holding the wire almost vertically in the Bunsen flame till it becomes red-hot.

(f) Take a sample of freshly collected soil on the wire loop and, raising the 'Experiment' lid carefully as in Fig. 2, scatter the soil over the agar so that the particles are widely separated.

(g) Open the aluminium foil packet of sterile soil (the soil has been heated to 120 °C for 90 min) and sterilize the wire loop once again in the Bunsen flame.

(h) Transfer a little of the *sterilized soil* into the 'Control' dish in the same way as before.

(i) The Petri dishes will now be incubated at 23-30 °C for two days or left in a cupboard at room temperature for a week.

(j) After the period of incubation, *do not remove the lids* but examine the contents of the dishes and answer the questions opposite.

(Questions 1-3 need to be answered while the dishes are available for study.)

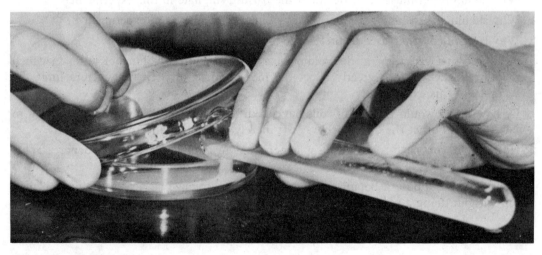

Fig. 1. Pouring the agar

Experiment 10. Discussion

1 Is there any kind of growth on the agar surface in **(a)** the 'Experiment' or **(b)** the 'Control'?

2 If there is, try to describe the different types of growth, e.g., white, yellow, brown, fluffy, slimy, branching, compact, and try to suggest whether the colonies are of bacteria or fungi.

3 Are all the colonies restricted to the vicinity of the soil particles or do some occur in the absence of soil particles?

   Suggest explanations for those colonies which have a soil particle at the centre and for those colonies which are not associated with visible particles.

4 Why can you be reasonably certain that the bacterial or fungal particles have come from the soil and have not arisen from bacteria or fungi already present on the glass dish, in the agar, on the wire loop or in the air when you lifted the lid of the dish?

5 Why did you not use your fingers to transfer the soil to the dish?

6 If colonies of micro-organisms appeared in the control dish, **(a)** how could this be explained and **(b)** why would it throw doubt on the origin of the colonies in the 'Experiment' dish?

7 The micro-organisms in your dish have fed on the nutrients in the agar jelly. What do you suppose is their normal source of food in the soil?

8 Suppose there were fifty different types of bacteria and fungi in your soil sample why do you think it unlikely that you would have obtained colonies of all these on the agar?

9 If you took samples of soil from successively deeper layers of soil, do you think that the micro-organisms would be **(a)** more or less numerous, **(b)** of the same types as are present in the upper layers?

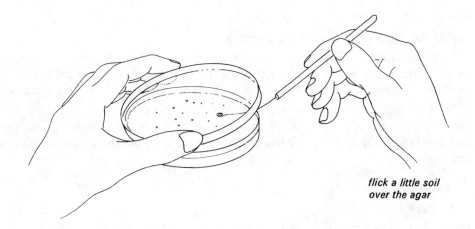

flick a little soil over the agar

lift the lid as little as possible

Fig. 2.

89

Experiment 11. Nitrifying bacteria in the soil

You are provided with two funnels plugged with glass wool and containing equal weights of freshly collected (A) and heat-treated (B) soil. You will also have available a reagent bottle containing diphenylamine in concentrated sulphuric acid which must be handled with care. Do not, at any stage, disturb the soil in the funnel.

(a) Support funnel A in a clamp or ring and place a small beaker under the stem leaving a large enough gap to interpose a spotting tile (Fig. 1).

(b) Place the clamp stand, funnel, beaker and spotting tile in a suitable tray.

(c) Fill a measuring cylinder with distilled water, pour 10 cm^3 from it on to the soil in funnel A, wetting the surface evenly, and wait for 10 seconds.

(d) If no water passes through the soil, pour another 5 cm^3 from the measuring cylinder into the soil and repeat this until water emerges from the funnel.

(e) When water starts to drip through, collect one drop in a cavity in the spotting tile. If soil particles are seen to be present, ignore this sample and collect another until you have a clear one.

CAUTION. **The diphenylamine you are going to use is dissolved in concentrated sulphuric acid. Keep the bottle in the tray at all times. Put the dropping pipette straight back into the bottle and not on the bench. Report any spillage and wash skin or clothing with much water. When washing the tile, take care not to splash the acid in the cavities over yourself or the surrounding bench. Use a gentle stream of water and hold the tile tilted away from you.**

(f) Place the tile in the tray and add three drops of diphenylamine to the drop of soil water. A blue colour indicates the presence of nitrates.

(g) Continue to add distilled water in 10 cm^3 volumes to the soil, collecting a drop of the filtrate after each addition and testing it for nitrates as above until no blue colour appears when diphenylamine is added. (About 50 or 60 cm^3 should be sufficient to wash all the nitrates from the soil.)

(h) Now use a graduated pipette to add 5 cm^3 dilute ammonium sulphate solution to the soil and test one of the final drops with diphenylamine.

(i) Repeat the whole procedure using funnel B.

(j) Take both funnels to the filter pump. Starting with funnel A, connect the stem to the pressure tubing and turn the tap on, a little at a time, to draw off excess liquid from the soil. Repeat this with funnel B. (About 15 seconds at the filter pump should be adequate.)

(k) Cover the mouths of the funnels with a Petri dish lid or aluminium foil. Write your initials on the cover with a spirit marker and leave for at least three days or, preferably, one week.

3 days to 1 week later

(l) Remove the covers from the funnels and pour distilled water, 5 cm^3 at a time, into each funnel in turn. In a spotting tile collect any one of the first five to ten drops to come through the soil and test these with three drops of diphenylamine, taking the same precautions as before.

Experiment 11. Discussion

1 What is the significance of the blue colour which appears when diphenylamine is added to soil water?

2 How do you interpret the fact that after 50-60 cm³ water has passed through the soil, no blue colour is given with diphenylamine?

3 When you tested the soil water with diphenylamine after an interval of several days, what result was obtained with water from funnels A and B and how do you interpret these results?

4 Why should heat treatment of soil in B cause it to give results different from A after the three- to seven-day period?

5 A critic of the experiment objects that the 50 cm³ water poured through the soil removed only some of the nitrate, and that after three to seven days soaking in water the residual nitrate is washed out and gives a blue colour with diphenylamine. What evidence have you to refute this?

6 The same critic, undaunted, says that heating the soil in B caused the nitrates to decompose and that this accounts for the failure of soil B to give a blue colour when tested at the end of several days. How can you counter this argument?

7 What do you think is the role of the ammonium sulphate in this experiment? Do you think you would have obtained similar results if ammonium sulphate had not been added? Explain.

8 How do you think that heavy and prolonged rainfall is likely to affect the nitrate content of a soil?

Fig. 1. Collecting a sample of soil water

Experiment 12. The mineral composition of soil

(a) Weigh out 50 g dry soil which has been sieved through a 10-mesh sieve and place the soil in a mortar.

(b) Grind the soil with a pestle for two or three minutes. This will help to break up the aggregates of particles without fracturing the particles themselves to any great extent.

(c) Tip the ground-up soil on to a sheet of paper.

(d) Partly fold the paper and pour the soil into a flat-sided medicine bottle.

(e) Fill the bottle with water from the tap up to the neck. Screw the cap on securely and shake the bottle vigorously for about 30 seconds to disperse the soil in the water.

(f) After shaking, stand the bottle upright on the bench and rock it very slightly from side to side to vibrate the soil as it settles down during the first minute. Allow the bottle to stand for 10-15 minutes.

(g) Copy the table given below into your notebook.

(h) When the soil has settled, examine it from the side and try to determine the boundaries between the following layers: coarse sand, fine sand, silt or clay. Mark the point where you think the boundaries occur with a spirit marker on the outside of the bottle and use a ruler to measure the depth of each layer in mm. Record these values in your table.

    (Bear in mind that the soil sample has already been sieved to remove particles greater than 2.5 mm (1/10 in).

(i) Answer questions 1-3 while you still have the bottle and soil before you to examine.

(j) If other samples of soil are available for comparison, empty the bottle into the container provided and wash it out. Then repeat the experiment with the fresh sample.

Particles	Sample 1: depth in mm	Sample 2: depth in mm
Coarse sand		
Fine sand		
Silt or clay		

vibrate the bottle

Experiment 12. Discussion

1 Are the boundaries between the different layers of particles quite distinct? Why is it unlikely, in most cases, that there will be distinct boundaries?

2 What kind of material, if any, is floating at the surface?

3 Is the water still cloudy? If so, what explanation can you offer for its appearance?

4 Large soil particles are not necessarily more dense than small ones but they sink more rapidly. Can you suggest why this should be so?

Mark the boundaries

Experiment 13. Estimation of the mineral composition of the soil

You are provided with two tall jars with lids; the jars have a mark on the side, 100 mm from the bottom.

(a) Weigh out 50 g dry soil which has been passed through a 10-mesh sieve.

(b) Place the soil in a mortar and grind gently with a pestle to break up the larger aggregates.

(c) Transfer the powdered soil to a piece of scrap paper and tip it into one of the jars.

READ **(d)** and **(e)** before continuing

(d) Pour 'Calgon' solution into the jar with the soil till it reaches the 100 mm mark. Replace the lid and shake the mixture for 3 minutes. (The 'Calgon' disperses the clay aggregates and keeps the clay particles in suspension.)

(e) As quickly as possible after the three minutes' shaking, remove the lid and pour the liquid into the second jar, leaving the sand in the bottom of the first jar.

(f) Wash the sand in jar 1 by adding successive portions of water (about 30 cm^3), swilling the sand and water about and pouring off the liquid into jar 2. Continue to do this until the liquid in jar 2 is up to the mark.

(g) Place the lid on jar 2 and shake the contents for about 30 seconds. Note the time and leave the jar to stand for five minutes.

(h) While waiting for the fine sand to settle in jar 2, tip the sand from jar 1 into a metal tray, adding water and swilling out the jar as necessary to transfer all the sediment to the tray. Tip the surplus water from the tray and place the tray on a tripod over a low Bunsen flame to dry the sand. (Keep an eye on the time for jar 2.)

(i) After jar 2 has stood for five minutes pour away the liquid and suspended clay leaving the fine sand behind.

(j) Transfer this sediment to a metal tray as before and tip off the surplus water but without losing too much of the solid.

(k) Heat the metal tray over a low Bunsen flame to dry the fine sand, moving the Bunsen about to avoid burning the organic matter.

(l) When both solids are dry, allow the trays to cool, scrape out the dry residue on to pieces of scrap paper, weigh the two fractions separately and record the weights in a table similar to that given below. Multiply by 2 to obtain the percentages and calculate the silt/clay fraction by subtraction.

Type of soil.............................. *Location*...............................			
Mineral particles	*Diameter of particles in* **mm**	*Weight*	*Percentage*
Coarse sand	2.0-0.2		
Fine sand	0.2-.02		
Silt	.02-.002		
Clay	below .002		

(To separate silt and clay by this method takes about 8 hours.)

94

Experiment 13.   Discussion

1 Explain why a sedimentation technique of this kind separates out particles of different sizes.

2 If the clay aggregates had not been dispersed by the 'Calgon' solution, why would the results obtained be less accurate?

3 The soil sample contained humus and other organic matter. How do you think the presence of humus affects your analysis of the mineral fractions?

4 From the appearance of the various fractions, which do you think contained the most organic matter? How would this affect the results?

5 The following table shows one example of the distribution of mineral particles in three types of soil.   Into which category of soils does your sample fall?

	Sandy loam	*Clay loam*	*Clay*
% coarse sand	65	30	1
% fine sand	20	30	9
% silt	5	20	25
% clay	10	20	65

(From *Principles of Plant Physiology*, James Bonner and Arthur W. Galston. W. H. Freeman & Company. Copyright © 1952.)

Pour the liquid into the second jar, leaving the sand at the bottom

Experiment 14.   Small animals in the soil

The apparatus used to extract the animals from the soil is illustrated in **Fig. 2.**

(a) To collect the soil sample, moisten the small can and push it into the soil until its upper rim is level with the soil surface. It may be necessary to push it down with your foot or to change to a different area if you encounter a large stone.

(b) Dig the can out with a trowel or garden fork and cut the soil level at the base of the can.

(c) Use the lid of the can to push the soil core out of the can into a container and take the sample back to the laboratory or store it in a polythene bag until it can be used.

(d) Place the flower pot on a sheet of paper and ensure that the gauze is in place in the bottom. Crumble the soil sample into the flower pot, breaking the soil into its natural crumbs but without compressing or smearing the crumbs.

(e) Use a graduated pipette to place 4 cm³ of the preserving liquid in the plastic Petri dish lid.

(f) Carry the flower pot and Petri dish to the part of the laboratory where they will stand for a week, i.e. within reach of an electricity outlet.

(g) Fit the lamp house into the top of the flower pot and place the flower pot on top of the Petri dish.

(h) Connect the lamp house to the electricity supply and switch on the current. Leave in position for at least five days.

(i) After five to seven days, switch off the lamp, lift the flower pot off the dish and search the dish for animals under the lowest power objective of the microscope as follows:

  (i) If there is still a lot of liquid in the dish, remove some of it with a fine pipette so that the floating animals do not swill about with the movements of the dish.

  (ii) Focus the microscope on a line scratched in the dish and, starting from one side, move the dish backwards and forwards so that you can examine the area between two lines (Fig. 1). Then move to the next line and so on across the dish, counting the different types of animal as you go. (See p. 98 for illustrations.)

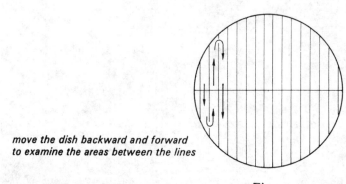

move the dish backward and forward to examine the areas between the lines

Fig. 1.

Experiment 14. Discussion

1 What factors do you think cause the animals to leave the soil and fall into the Petri dish?

2 Why do you think it unlikely that *all* the animals present in the soil sample will be driven out by this method?

3 Assuming that only a proportion of the animals actually in the soil are driven out, do you think you will obtain a *representative* sample? Explain.

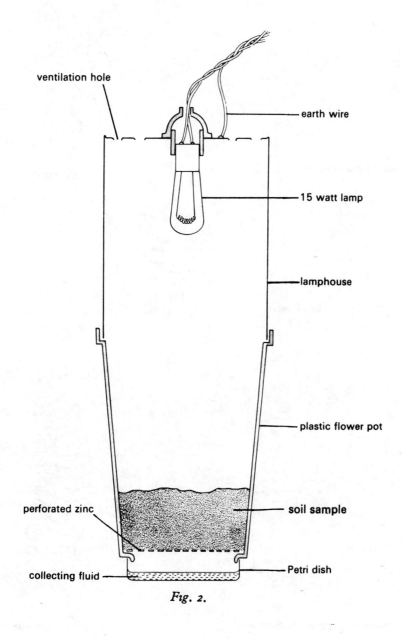

Fig. 2.

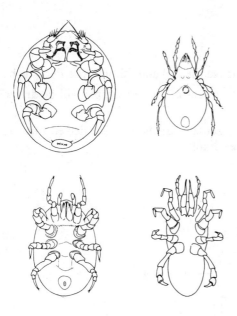

Some different types of mites

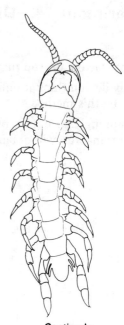

Centipede

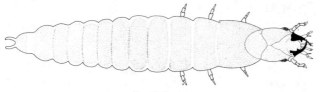

Beetle larva

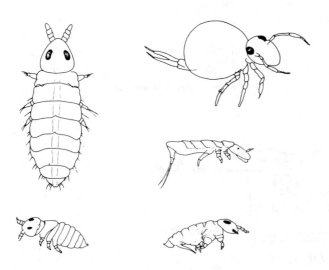

Some different types of springtails (Collembola)

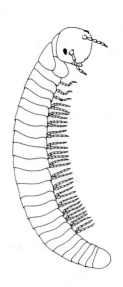

Millipede

4 Photosynthesis

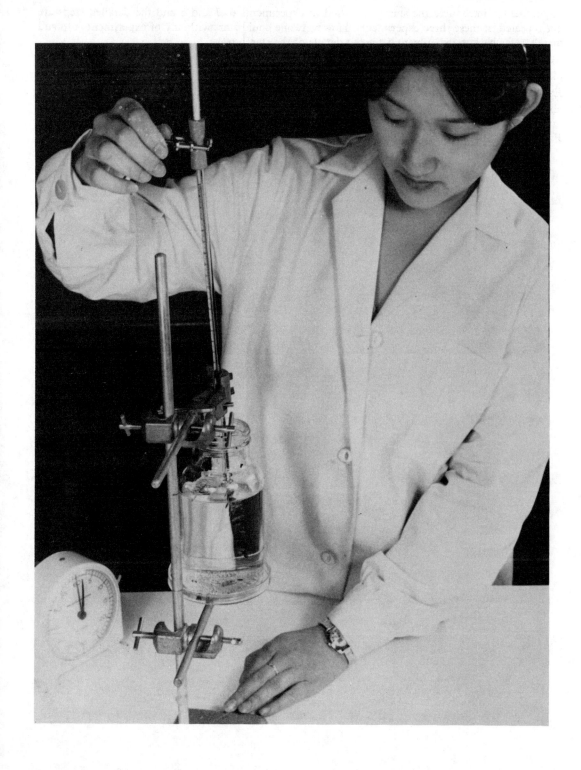

Introduction

The order in which the experiments are presented is not intended to represent any particular sequence in the development of ideas about photosynthesis but experiment 3 must precede 3a-c since it gives the instructions and practice necessary for using the microburette required in 3a, b and c. This series of experiments is best suited to 5th and 6th form students.

Experiment 5 introduces the starch test used in experiments 6, 7 and 8 and the detailed steps are not repeated in these three experiments. However, one could start with any of experiments 6-8 and merely refer back to experiment 5 for the details of the starch test.

Experiments 4a and 4b are alternatives, demonstrating the same effect with either a land or a water plant.

Experiments 11a and 11b are not alternatives but 11b can be tackled without first attempting 11a. Experiments 1, 3, 5, 11a and 12 are intended to work in an hour or less; experiments 6 and 7 will give results in an hour but are more certain to succeed if the experiment continues for several hours. Experiments 2, 8, 9 and 11b need 3-7 days; experiment 10 needs 2 weeks for satisfactory results.

Contents

Experiment 1. Production of gas by pond-weed

(a) Fill a beaker or glass jar with tap-water and add about 5 cm³ (20 mm in a test-tube) saturated sodium hydrogencarbonate solution.

(b) Select a pond-weed shoot between 5 and 10 cm long.

(c) Take a small paper clip, prize it slightly open and slide it over the pond-weed shoot a short distance behind the growing point. This is to hold the weed down in the water.

(d) Place the pond-weed in the jar so that it floats vertically with the cut end of the stem uppermost but still entirely below the water.

(e) Switch on the bench lamp and bring it close to the jar.

(f) After a minute or two, bubbles should appear from the cut end of the stem. If they do not, obtain a fresh piece of plant material and try again.

(g) When the bubbles are appearing with regularity, switch the lamp off and observe any changes in the production of bubbles.

(h) Switch the lamp on and place it about 25 cm from the pond-weed. Try to count the number of bubbles appearing in a minute. Now move the lamp to about 10 cm from the plant and again try to count the number of bubbles.

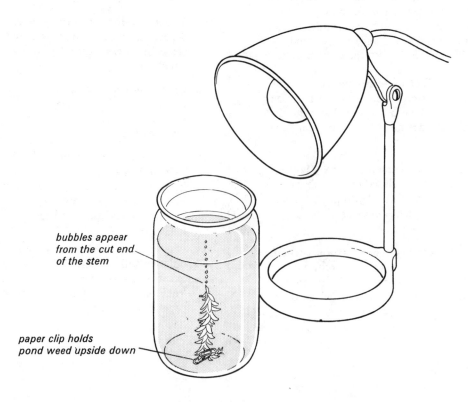

bubbles appear
from the cut end
of the stem

paper clip holds
pond weed upside down

Experiment 1. Discussion

1   Although the bubbles are seen to escape from the cut end of the stem, which parts of the pond-weed could be producing the gas?

2   In what way does the production of gas appear to be related to the intensity of light reaching the plant?

3   Suppose that putting the lamp at 10 cm caused twice as many bubbles to appear per minute as when it was at 25 cm. Why would you *not* be justified in saying that the production of gas by the plant had doubled?

4   (a)   Is there any evidence to suggest what gas or gases might be present in the bubbles?
    (b)   Bearing in mind the composition of atmospheric air and the composition of water, what gases might be present in the bubbles?

5   What was the point of adding sodium hydrogencarbonate to the water at the beginning of the experiment?

Slide the paper clip over the pond weed near the tip

Experiment 2.   Collecting the gas evolved by pond-weed

You are provided with a glass jar having a lid through which a test-tube can pass, a rubber band and a cork with a wide hole in it.

(a) Pass the test-tube up through the hole in the lid and secure it with the elastic band as shown in Fig. 3.

(b) Write your initials on the jar and fill it nearly to the top with tap-water. Add about 5 cm³ sodium hydrogencarbonate solution (about 20 mm in a test-tube).

(c) Collect about 10 pieces of pond-weed up to 10 cm long. Arrange the shoots parallel to each other and trim the ends of the stems with a razor blade.

(d) Fill the test-tube with water to the top of the cork. Push the cut ends of the pond-weed stems through the hole in the cork so that they are held firmly but without crushing them (Fig. 1).

(e) Hold the test-tube horizontally with the cork over the edge of the jar (Fig. 2) and then turn it upside down so that the pond weed enters the jar, the lid fits over the opening and the test-tube is held by the elastic band (Fig. 3).

(f) Little or no air should enter during this operation but if, for some reason, more than about 10 mm air gets into the tube, the experiment should be set up again. Make sure that the cork is under the water in the jar.

(g) Repeat the whole operation from **(a)** to set up an identical experiment but cover the jar with aluminium foil or some opaque material to exclude all light. Place both jars in a position where they can receive maximum daylight or artificial light (Fig. 4).

TESTING THE GAS. READ ALL THIS SECTION BEFORE PROCEEDING.

(h) When either of the test-tubes is more than half full of gas, remove the elastic band and lid, lift the tube and weed out of the jar and turn the tube the right way up. Remove the cork and pond-weed and quickly close the mouth of the tube with your thumb (Fig. 5).

(i) Light a wooden splint and when it is burning well, blow it out so that the end continues to glow red.

(j) Remove your thumb from the test-tube and at the same time insert the glowing end of the splint into the tube. (By closing the tube again, blowing out the splint and re-inserting it, you may be able to repeat the test several times.)

Experiment 2.   Discussion

1 What happened to the glowing splint when it was placed in the test-tube?
2 What gas usually causes this reaction?
3 The test does not prove that *only* this gas is present in the test-tube. What other gases might be present?
4 What evidence do you have which suggests that the production of this gas depends on light reaching the plant?
5 What evidence is there to suggest that the gas in the test-tube came from the pond-weed and not directly from the water in the jar?
6 How could you eliminate this last possibility?

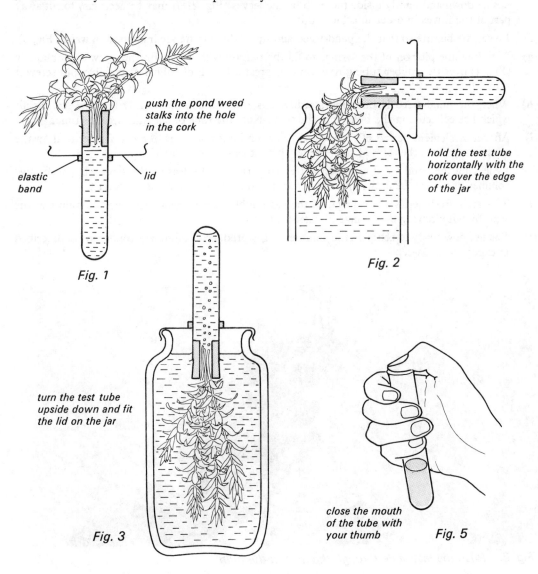

push the pond weed stalks into the hole in the cork

elastic band

lid

Fig. 1

hold the test tube horizontally with the cork over the edge of the jar

Fig. 2

turn the test tube upside down and fit the lid on the jar

Fig. 3

close the mouth of the tube with your thumb

Fig. 5

Experiment 3.   Conditions affecting the production of oxygen

(a)   Fill a large beaker or jar with tap-water up to the mark.

(b)   Fit a clamp stand with a ring, gauze and clamp as shown in Fig. 1 so that the beaker is on a level with the bulb in a bench lamp.

(c)   Use the clamp to grip the stem of the burette with the bulb clear of the water.

(d)   Collect 3 or 4 pieces of pond-weed, 5-10 cm long, and cut the ends with a razor blade mid-way between two nodes. Place the shoots upside down in the beaker, switch on the lamp to be used for the experiment and place it about 15 cm from the beaker. When bubbles appear from the stems, select a shoot which is bubbling rapidly and regularly and return the rest.

(e)   Push the last node of the selected shoot into the hole at the base of the burette so that it is held upside down and lightly inside the bulb by the leaves (Fig. 2). It may be necessary to cut away part of the leaves in order to achieve this.

(f)   Lower the burette so that the pond-weed and tip of the burette are immersed in water (Fig. 3).

(g)   Withdraw the plunger of the syringe to fill the burette with water. Fit the screw clip obliquely (Fig. 4) over the rubber tubing to enclose as great a length of rubber as possible and screw it shut.

(h)   When the pond-weed is bubbling steadily, unscrew the clip sufficiently to draw out the air which has collected in the bulb, into the stem. Note the time or start the seconds timer.

(i)   After a convenient period of time, e.g. 1, 2 or 3 minutes, unscrew the clip farther to draw the collected gas into the stem of the burette where it can be measured.

(j)   By leaving the pond-weed to bubble for equal periods and releasing the screw clip, successive volumes of gas can be drawn into the graduated stem and the readings averaged.

(k)   To start a fresh series of readings, all the gas bubbles can be removed from the bulb and the stem by withdrawing the syringe plunger as in **(g).**

(l)   You are now ready to compare the volumes of gas produced in different conditions as described in experiments 3a-c.

Fig. 2.   Push the last node through the hole in the bulb

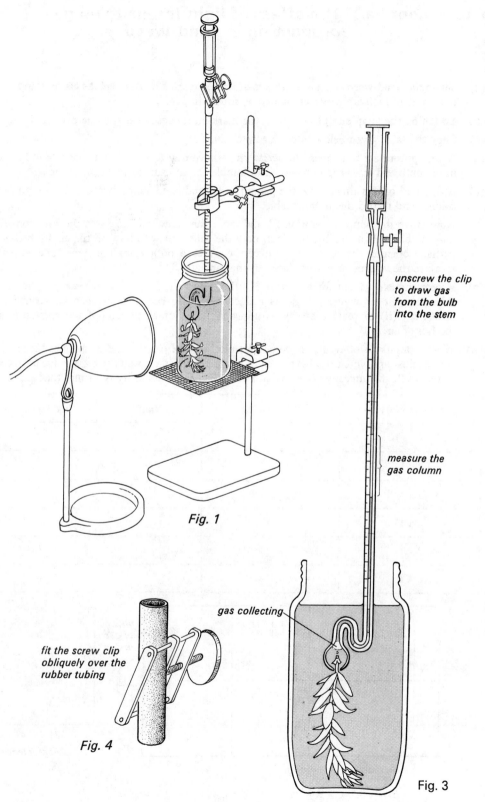

unscrew the clip
to draw gas
from the bulb
into the stem

measure the
gas column

Fig. 1

gas collecting

fit the screw clip
obliquely over the
rubber tubing

Fig. 4

Fig. 3

Experiment 3a.   The effect of light intensity on gas production in pond-weed

(a)   Set up the pond-weed and gas burette as described on p. 106. Arrange the clamp stand at one end of a sheet of foolscap paper graduated as shown in Fig. 1.

(b)   Switch on the lamp and place it with the mark on its base over the line marked I (Fig. 2).

(c)   Copy the table given below into your notebook.

(d)   At an appropriate time open the screw clip sufficiently to remove any gas which has collected in the bulb and leave the pond-weed to bubble for an exact number of minutes.

(e)   At the end of this time, draw the gas into the graduated stem, measure the length of the gas column and record this in your table.

(f)   Now move the lamp to position 2I and close the screw clip to force the gas from the stem into the bulb. At a suitable moment, note the time and withdraw all the air from the bulb by partially opening the screw clip. Leave the weed to bubble for the same time as before and measure the volume of gas produced in this time.

(g)   Repeat the experiment for distances 3I, 4I, 5I and 6I. It may be necessary **(i)** to reduce the time period over which the gas is collected if more accumulates than can conveniently be measured, **(ii)** to remove air by withdrawing the syringe plunger if the apparatus becomes too full of air.

(h)   When your table of results is complete, plot a graph of volume of gas produced in a fixed time against light intensity, bearing in mind that the distances on the paper have been calculated so that at 2I the intensity is twice that at I and at 3I the intensity is 3 times that at I and so on.

Light intensity	Volume of gas	Time	Volume per minute
I			
2 x I			
3 x I			
4 x I			
5 x I			
6 x I			

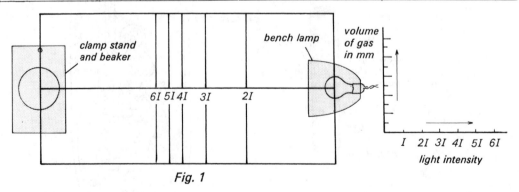

Fig. 1

Experiment 3a.   Discussion

1 What general effect did increasing light intensity have on gas production?

2 How could you explain the result in terms of photosynthesis?

3 In this experiment why is the rate of gas production not necessarily related to the rate of photosynthesis?

4 What additional information would you need before you could assume that a doubling of gas production also meant a doubling in the rate of photosynthesis?

5 In this experiment, the light intensity was progressively increased in equal steps throughout. Did the production of gas also increase in equal steps over the whole range of light intensities?

6 If the light intensity continued to increase beyond the range of your experiment, would you expect the gas production to continue increasing in proportion? Explain your answer.

7 What condition other than light intensity might have been altered inadvertently during the course of this experiment? How would you expect changes in this condition to affect the results?

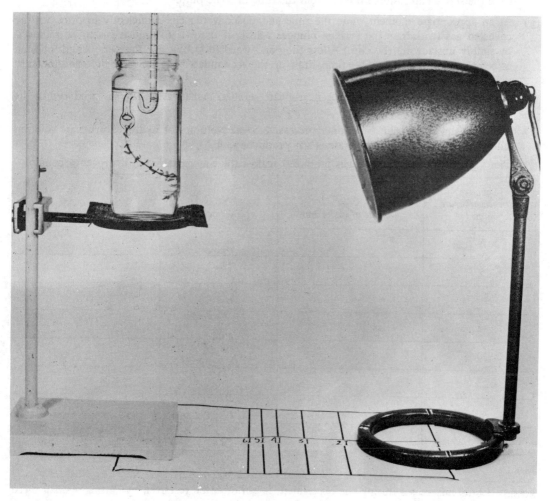

Fig. 2

Experiment 3b.   The effect of carbon dioxide concentration on gas production

(a)   Set up the apparatus as described on p. 106. Select a vigorously bubbling piece of pond-weed and insert it in the microburette.

(b)   Place the bench lamp 15 cm from the beaker and do not move it during the course of the experiment.

(c)   Copy the table given below into your notebook.

(d)   At an appropriate moment, note the time and unscrew the clip sufficiently to draw any gas which has accumulated in the bulb, up into the tube. Leave the pond-weed to continue bubbling for an exact number of minutes.

(e)   At the end of this time, draw the collected gas into the graduated tube, measure the length of the gas column and record this in your table.

(f)   Now add 5 cm³ sodium hydrogencarbonate solution to the jar from a graduated pipette and close the screw clip to force the air from the tube into the bulb.

(g)   At an appropriate moment, note the time and unscrew the clip sufficiently to draw up the collected gas (withdraw the syringe plunger a little if there is too much gas to be removed by simply unscrewing the clip.) Allow the pond-weed to bubble for the same length of time as before and at the end of this time, draw up the accumulated gas from the bulb and measure its length in the graduated tube.

(h)   Add another 5 cm³ sodium hydrogencarbonate solution and measure the gas produced in the time you have selected.

(i)   Continue the experiment by adding successive 5 cm³ volumes of hydrogencarbonate solution until there is no change in the rate of gas production.

(j)   Plot a graph of the volume of gas produced against the concentration of hydrogencarbonate.

Volume of hydrogencarbonate in cm³	Volume of gas in mm	Time	Volume per minute
0			
5			
10			
15			
20			
25			
30			
35			
40			

NOTE.   If bubbling ceases during the experiment it can often be restarted by cutting a small piece from the stem without having to obtain a fresh shoot and start the experiment all over again.

Experiment 3b.　Discussion

If you have already answered the questions on experiment 3a, questions 3-5 can be ignored.

1　What general effect did increasing the concentration of hydrogencarbonate have on gas production?

2　How could you explain these results in terms of photosynthesis?

3　In this experiment, why is the rate of gas production not necessarily related to the rate of photosynthesis?

4　What additional information would you need before you could assume that a doubling of the gas production also meant a doubling in the rate of photosynthesis?

5　In this experiment, the hydrogencarbonate concentration was progressively increased in equal steps throughout. Did the production of gas also increase in equal steps over the whole range of hydrogencarbonate concentrations?

6　If the hydrogencarbonate concentration continued to increase beyond the range of your experiment, would you expect the gas production to continue increasing in proportion? Explain your answer.

7　What conditions other than the hydrogencarbonate concentration might have been altered during the experiment? How might these alterations affect the rate of gas production?

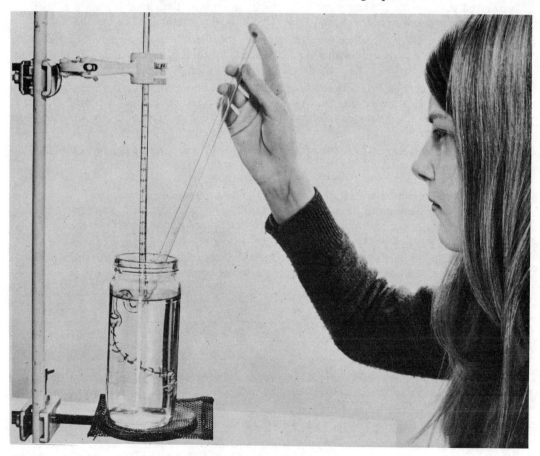

Add 5 cm³ sodium hydrogencarbonate solution

Experiment 3c. The effect of temperature on gas production

(a) Set up the apparatus with the pond-weed as described on p. 106 and check that it is within easy reach of a Bunsen burner. For this experiment a heat-resisting beaker must be used. A glass jar will not do.
Choose a shoot which is bubbling steadily but at a moderate rather than a rapid rate.

(b) Copy the table given below into your notebook.

(c) Switch on the bench lamp, place it 15 cm from the beaker and do not move it during the experiment.

(d) Place a thermometer in the beaker, resting it in the lip of the beaker so that it does not roll about but can still be read without removing it. Add two or three pieces of ice to the beaker and stir the water carefully with the thermometer. Continue adding ice until the temperature falls to about 15 °C.

(e) At a convenient moment, note the time and unscrew the clip to remove any gas which has collected in the bulb.

(f) Leave the pond-weed to produce bubbles for an exact number of minutes and then draw the collected gas into the capillary tube by partially unscrewing the clip. Measure the length of this gas column and record it in your table. Check the temperature and record this also.

(g) Close the screw clip to expel air from the capillary tube into the bulb. Remove any remaining ice and heat the water gently with a small Bunsen flame until it reaches about 20 °C. Remove the flame.

(h) At an appropriate moment, note the time and draw out all the air from the bulb by partially unscrewing the clip. Leave the plant to produce gas for the same time as before. Draw the collected gas into the graduated capillary, measure its length and record it in your table. Note and record the temperature.

(i) Repeat the experiment at intervals of about 5 or 10 °C according to the time available, up to 50 °C.

NOTE. (i) For temperatures above 35 °C it is best to take three successive readings of gas volumes at each temperature and average the result. This can be done by simply unscrewing the clip at the end of each time interval for three successive intervals giving, finally, three separate columns of gas in the graduated tube.

(ii) If bubbles start to appear from parts of the shoot other than the cut stem, a fresh cut should be made across the stem inserted in the burette.

Temperature	Volume of gas	Time	Volume per minute

Experiment 3c. Discussion

1 What effect, in general, did a rise in temperature have on the rate of gas production?

2 Why should a change in temperature affect the rate of photosynthesis?

3 In what way, other than its effect on photosynthesis, could a rise in temperature affect the volume of gas escaping from the pond-weed?

4 In this experiment, why is it that an increase in the volume of gas produced is not necessarily a reliable indication of an increase in the rate of photosynthesis (two reasons)?

5 Was the change in the rate of gas production consistent over the whole range of increasing temperature? If not, offer an explanation.

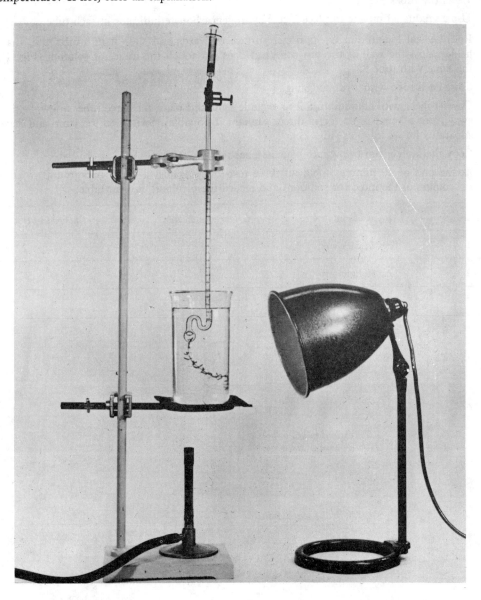

Experiment 4a. Gaseous exchange in leaves

The experiment depends on the use of hydrogencarbonate indicator, a pH indicator, containing the dyes cresol red and thymol blue in a solution of sodium hydrogencarbonate. This pH indicator is in equilibrium with atmospheric carbon dioxide, i.e. its orange colour when you receive it indicates the acidity of the atmosphere due to carbon dioxide. Carbon dioxide is an acid gas.

Increase in acidity (fall in pH) turns the indicator yellow while decrease in acidity (rise in pH) turns it first red and eventually purple.

(a) Wash three test-tubes in tap-water. Rinse them with distilled water and finally rinse them with the hydrogencarbonate indicator itself.

(b) Label the tubes 1-3.

(c) Use a graduated pipette to place 2 cm³ hydrogencarbonate indicator in each tube.

(d) Roll the leaf longitudinally, its upper surface outwards and slide it into tube 1 so that it is held against the wall of the test-tube and does not touch the indicator solution (Fig. 1). Do the same with tube 2.

(e) Close each tube with a rubber bung.

(f) Cover tube 1 with aluminium foil to exclude light and place the three tubes a few centimetres away from a bench lamp (or in direct sunlight if possible). Switch on the lamp and leave the apparatus for 40 minutes (Fig. 2).

(g) Copy the table given below into your notebook.

(h) At the end of 40 minutes, hold all three test-tubes against a white background to compare the colours of the indicator solutions and record these colours in your table.

Tube	Conditions	Colour of indicator	Change in pH
1	Leaf in darkness		
2	Leaf in light		
3	No leaf		

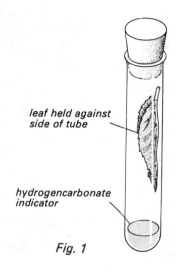

leaf held against side of tube

hydrogencarbonate indicator

Fig. 1

Experiment 4a.   Discussion

Read the introductory paragraphs to the experiment once again.

1 Although the hydrogencarbonate indicator solution is a pH indicator, i.e. its colour depends on its acidity, this experiment assumes that its changes of colour depend entirely on changes in the carbon dioxide content of the air. Explain why this is a reasonable assumption.

2 (a) What change in pH is suggested by the indicator becoming more yellow?
  (b) What change in the composition of the air in the test-tube is most likely to cause such a change of pH?

3 (a) What change in pH is suggested by the indicator becoming more red or purple?
  (b) What change in the composition of the air in the test-tube is most likely to cause such a change in pH?

4 What was the purpose of setting up tube 3?

5 What effect did the presence or absence of light have on the concentration of carbon dioxide in tubes 1 and 2?

6 If you know something about the processes in leaves which lead to gaseous exchange, explain the difference in results in test-tubes 1 and 2.

Fig. 2

Experiment 4b.   Gaseous exchange in pond-weed

The experiment depends on the use of hydrogencarbonate indicator, a pH indicator containing the dyes cresol red and thymol blue in a solution of sodium hydrogencarbonate. This pH indicator is in equilibrium with atmospheric carbon dioxide, i.e. its orange colour when you receive it indicates the acidity of the atmosphere due to carbon dioxide.

Increase in acidity (fall in pH) turns the indicator yellow while decrease in acidity (rise in pH) turns it first red and eventually purple.

(a)   Wash three test-tubes in tap-water. Rinse them with distilled water and finally rinse them with the hydrogencarbonate indicator itself.

(b)   Label the tubes 1-3.

(c)   Select 4 equivalent shoots of pond-weed about 50 mm long and place 2 shoots in each of tubes 1 and 2.

(d)   Pour hydrogencarbonate indicator into all three tubes to the same level and sufficient to cover the pond-weed in tubes 1 and 2 (Fig. 1).

(e)   Close each tube with a rubber bung.

(f)   Cover tube 1 with aluminium foil to exclude light and place the three tubes a few centimetres away from a light source or in direct sunlight if possible. Leave the tubes in the light for 40 minutes (Fig. 2).

(g)   Copy the table given below into your notebook.

(h)   At the end of 40 minutes, hold all three test-tubes against a white background to compare the colours of the indicator solutions and record these colours in your table.

Tube	Conditions	Colour of indicator	Change in pH
1	Pond-weed in darkness		
2	Pond-weed in light		
3	No pond-weed		

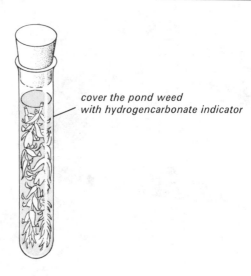

cover the pond weed
with hydrogencarbonate indicator

Fig. 1

Experiment 4b. Discussion

Read the introductory paragraphs to the experiment once again.

1   Although the hydrogencarbonate indicator solution is a pH indicator, i.e. its colour depends on its acidity, this experiment assumes that its changes of colour depend entirely on changes in the carbon dioxide content of the air. Explain why this is a reasonable assumption.

2   (a) What change in pH is suggested by the indicator becoming more yellow?

  (b) What change in the composition of the water in the test-tube is most likely to cause such a change of pH?

3   (a) What change in pH is suggested by the indicator becoming more red or purple?

  (b) What change in the composition of the water in the test-tube is most likely to cause such a change in pH?

4   What was the purpose of setting up tube 3?

5   What effect did the presence or absence of light have on the concentration of carbon dioxide in tubes 1 and 2?

6   If you know something about the processes in pond-weed which lead to gaseous exchange, explain the difference in results in test-tubes 1 and 2.

Fig. 2

117

Experiment 5.   Testing a leaf for starch

(a)   Half fill a 250 cm³ beaker or tin can with water (hot water if available) and place it on a tripod over a Bunsen burner. Heat the water till it boils and then turn down the Bunsen flame sufficiently to keep the water at boiling point.

(b)   Hold the leaf in forceps and plunge it into the boiling water for 5 seconds. This will kill the cells, arrest all chemical reactions and make the leaf permeable to alcohol and iodine solution later on.

(c)   With the forceps, push the leaf carefully to the bottom of a test-tube and cover it with methylated spirit (Fig. 1).

(d)   TURN OUT THE BUNSEN BURNER.

(e)   Place the test-tube in the hot water and leave it for 5 minutes. The alcohol will boil and dissolve out the chlorophyll in the leaf (Fig. 2).

(f)   Use a test-tube holder to remove the test-tube from the water bath and tip the green alcoholic solution into the receptacle for waste alcohol but take care not to tip the leaf out as well.
If the leaf is white or very pale green, go on to **(g)**.
If there is still a good deal of chlorophyll left in the leaf, boil it for a further 5 minutes with a fresh supply of alcohol, using the hot water bath. If it is necessary to relight the Bunsen to heat the water to boiling point, remove the test-tube and do not replace it until the Bunsen flame is extinguished.

(g)   Fill the test-tube with cold water and the leaf will probably float to the top.
    1.   Tough leaves (e.g. *Tradescantia*). Hold the leaf stalk with forceps and dip it into the hot water in the water bath for 2-3 seconds. Spread it flat on a tile or Petri dish lid with the aid of a little cold water.
    2.   **Soft leaves (e.g. busy Lizzie). Use forceps to place the leaf on a tile or back of a Petri dish lid and, holding the leaf stalk firmly against the tile or lid, let a fine trickle of water from the cold tap run over it to wash away the alcohol.**

(h)   If necessary, use the forceps to spread the leaf quite flat on the tile or lid. Using a dropping pipette cover the leaf with iodine solution for one minute.

(i)   Take the leaf to the sink and holding it on the tile or lid, wash away the iodine solution with a fine trickle of cold water.

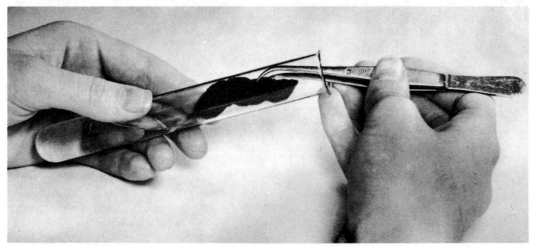

Fig. 1.   Push the leaf to the bottom of the tube

Experiment 5. Discussion

1 What is the reason for extracting the chlorophyll from the leaf?
2 If one is testing a leaf for the products of photosynthesis, why is it necessary to arrest all chemical reactions after detaching it?
3 For what substance is iodine a test? What result do you see if this substance is present?
4 What was the colour of the leaf (a) immediately before adding iodine, (b) after adding iodine?
5 How do you interpret this change?
6 In subsequent experiments we are going to claim that the presence of starch in a leaf is strong evidence that photosynthesis has taken place. Why is the result in the present experiment *not* good evidence of this?
7 What products of photosynthesis might be present which are not revealed by this test?
8 Assuming that the leaf has, in fact, been photosynthesizing and has produced starch, do your results indicate whether starch is the first product of photosynthesis to be formed?

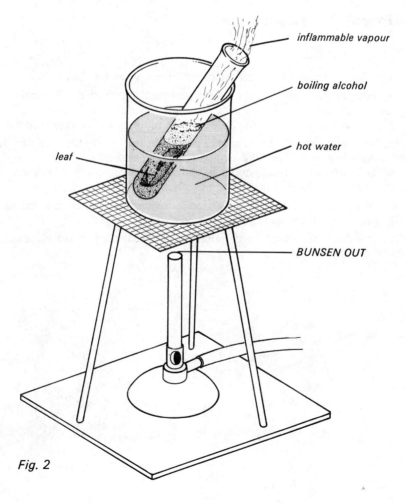

Fig. 2

Experiment 6.   The need for light in photosynthesis

You are provided with a potted plant which has been kept in darkness for at least 48 hours so that any starch in its leaves has been converted to sugar and removed.

(a)   Water the plant if the soil appears to be dry.

(b)   DO NOT REMOVE ANY LEAVES but select one at the top of the shoot, preferably one held out from the stem nearly horizontally.

(c)   Round this leaf wrap a strip of aluminium foil. Press it close to the surface so that this part of the leaf cannot receive any light at all. Put your initials on the foil with a spirit marker.

(d)   Place the plant under the light source so that the leaf is directly under the fluorescent tube. When the whole class has prepared the plants, the tube can be lowered to within a centimetre or two of the top leaves.

(e)   After a period of time (40-50 minutes is the minimum), detach the leaf with the foil and test it for starch as described on p. 118.

Experiment 6.   Discussion

1   Describe the appearance of the leaf after it had been tested for starch.

2   **(a)**   What is the significance of the colours and their distribution in the leaf?
   (b)   How could this be explained in terms of photosynthesis and light?
   (c)   Suggest at least one other way in which the results could be explained.

3   **(a)**   Apart from cutting off the light from part of the leaf, what other effect might the aluminium foil have had on the leaf which could influence the results?
   (b)   What control experiments could be conducted which would check whether these effects had influenced the results.

4   In this experiment, the plant was not tested at the beginning to ensure that its leaves were free from starch. Why was this precaution omitted?

5   Explain how the method used for destarching (not decolourizing) the plant tends to assume the results of this experiment.

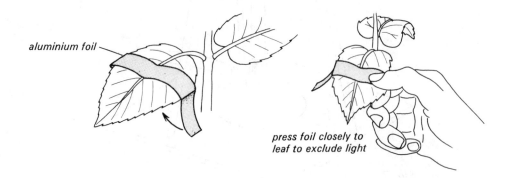

aluminium foil

press foil closely to
leaf to exclude light

Experiment 7. The need for chlorophyll

You are provided with a potted plant having variegated leaves. This plant has been in darkness for 2 days so that starch has been removed from the leaves.

(a) Detach a leaf from near the top of the plant and test it as described on p. 118. If any blue colour appears, the plant is unsuitable for the experiment.

(b) Water the plant if the soil looks dry.

(c) Place the plant in a situation where the leaf can receive sunlight or arrange the plant with the leaf nearly touching a fluorescent tube and leave it for at least 45 minutes.

(d) Detach a leaf from near the top of the plant and make a drawing in your notebook to show the outline and the areas which contain chlorophyll.

(e) Test this leaf for starch as described on p. 118.

(f) Alongside your first drawing of the leaf, make a second drawing after testing the leaf with iodine, to show its outline and the distribution of the blue colour due to starch and the brown colour due simply to staining with iodine.

Experiment 7. Discussion

1 What was the relationship between the distribution of chlorophyll in your leaf and the distribution of starch revealed by iodine?

2 How can you explain this *distribution* of starch in terms of photosynthesis?

3 What alternative explanations could there be for any correspondence between the distribution of chlorophyll and the distribution of starch?

4 Why was it necessary to draw the leaf before testing it for starch?

5 Why was there no need for a separate control experiment?

6 How do you know that there was no starch present in the leaf at the beginning of the experiment? Are you satisfied with this evidence?

Variegated Tradescantia cuttings   *Fig. 4*

Experiment 8. The need for carbon dioxide

You are provided with two potted plants which have been in darkness for 48 hours so that any starch present in their leaves has been removed.

(a) Label the pots A and B.

(b) Label two test-tubes A and B.

(c) Detach a leaf from near the top of each plant and test each one as described on p. 118 to ensure that no starch is present. If either leaf gives a blue colour the plant should not be used for the experiment.

(d) Water the plant if the soil seems dry.

(e) In a small container place about 20 g soda-lime. In a similar container pour about 10 cm^3 (40 mm in a test-tube) saturated sodium hydrogencarbonate solution. Place the container of soda-lime on the soil in pot A and the sodium hydrogencarbonate container in pot B (Fig. 1).

(f) Cover each plant with a transparent plastic bag, after checking it for holes. Secure the bags round the pots with elastic bands and mark the pots with your initials.

(g) Place both plants where they can receive daylight or artificial light for several hours (Fig. 2).

(h) Label two test-tubes A and B.

(i) After several hours of illumination remove the plastic bags from the plants and detach a leaf from near the top of each plant. Place the leaf in the correct test-tube to identify it.

(j) Test each leaf for starch as described on p. 118.

NOTE. Soda-lime absorbs carbon dioxide; sodium hydrogencarbonate solution decomposes slowly to give carbon dioxide.

Fig. 2

Experiment 8.   Discussion

1   Record your results in a table similar to the one below.

	A (no CO³)		B (control)	
	Colour	Interpretation	Colour	Interpretation
Leaf tested for starch at beginning of experiment				
Leaf tested for starch after ... hours light				

2   Assuming that the accumulation of starch in a leaf is evidence of photosynthesis make a general statement about photosynthesis and carbon dioxide.

3   In this experiment, what abnormal conditions other than removal of carbon dioxide might have affected the plant?

4   How far did the control (a) succeed, (b) fail, in showing that these other effects were not influencing photosynthesis?

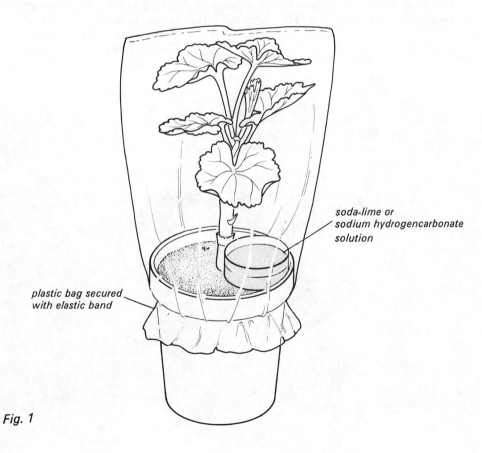

soda-lime or
sodium hydrogencarbonate
solution

plastic bag secured
with elastic band

Fig. 1

Experiment 9. Under what conditions can leaves produce starch in darkness?

(a) Copy the table given below into your notebook.

(b) Label two small beakers or glass jars A and B and add your initials.

(c) In A, place enough 5% glucose solution to half fill it.
In B, place an equivalent volume of tap-water.

(d) Cut a leaf from a plant which has been destarched by having been in darkness for 48 hours.

(e) Cut the leaf blade into 4 pieces of about equal area and small enough to fit easily in the beakers or jars.

(f) Test one piece of leaf for starch as described on p. 118. If it gives a blue colour, the leaf is unsuitable for the experiment. Further leaves should be selected and tested until one is found which is relatively free from starch.

(g) Float one piece of leaf blade, the right way up, on the surface of the glucose solution in jar A. Push a pin through another piece of leaf and sink it in the glucose solution in jar A. The pin is simply to hold it down in the liquid.
Float the remaining leaf portion on the surface of the water in B.

(h) Put both jars in a dark cupboard or under a cardboard box.

(i) After a period of 3-7 days, label three test-tubes A1, A2 and B and test the leaf portions separately for starch as described on p. 118. Enter the results in your notebook.

		Conditions	Result of test with iodine	Interpretation
		Leaf from destarched plant		
A1		Leaf floating on 5% glucose solution		
A2		Leaf submerged in 5% glucose solution		
B		Leaf floating on water		

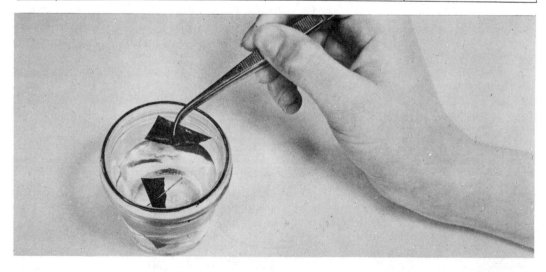

Experiment 9.   Discussion

1   Under what conditions was the leaf able to produce starch in the absence of light?

2   Suggest explanations to account for the leaf's failure to produce starch in the other conditions.

3   What additional control experiments would help to support your explanations in 2?

4   In experiments 6-8 it was assumed that the accumulation of starch in a leaf was evidence of photosynthesis. Discuss the validity of this assumption in the light of the results obtained from experiment 9.

5   Under what conditions and in what parts of a green plant would you expect (a) sugar to be converted to starch and (b) starch to be converted to sugar?

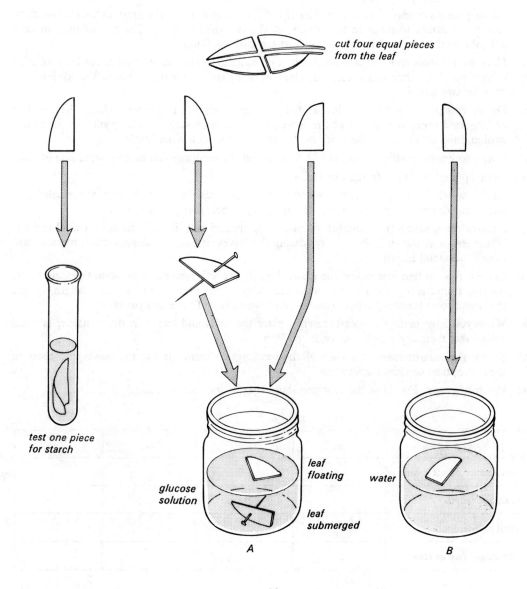

cut four equal pieces from the leaf

test one piece for starch

glucose solution

leaf floating

leaf submerged

water

A

B

Experiment 10. The need for mineral elements

(a) Label four test-tubes as follows: $+$, Ca, N, $-$.

(b) Fill each tube to within about 20 mm of the top, with the appropriate water culture.
   ($+$) Solution containing all mineral elements thought to be needed for healthy growth
   (Ca) Solution containing all mineral elements as in $+$ except for calcium
   (N) Solution as in $+$ but lacking nitrate
   ($-$) Distilled water, i.e. no mineral elements at all

(c) Select four seedlings which appear to be at the same stage of development.

(d) If there is a wide difference between the development of the root systems, reduce all systems to the same number and approximate length of root as in the least developed.

(e) Leaving the shoot and roots free, roll a strip of cotton wool round the grain to hold the seedling lightly but firmly in place in the mouth of the test-tube (Fig. 1). Place a seedling in each test-tube so that the root is well covered by the culture solution.

(f) Mark your initials and the date on the rack or container provided and place the four tubes in a position where they can receive daylight or artificial illumination. Leave the seedlings to grow for two weeks.

(g) During this period of time the levels of the solutions will fall in the test-tubes. They need to be inspected every two days and the level restored if necessary. Carefully remove the cotton wool and top up the test-tubes with DISTILLED WATER from a wash-bottle.

(h) After two weeks transfer the tubes to a rack so that the seedlings can be compared side by side.

(i) Draw up a table similar to the one below.

(j) Study the whole group of seedlings and note in your table any which show abnormalities of leaf colour or shape, e.g. dead areas, discoloured patches, pale green colour.

(k) Remove the seedling from the full culture ($+$), unwind and discard the cotton wool and cut off the leaves as shown in Fig. 2. By placing the leaves end to end along a ruler, measure and record their total length.

(l) Cut the root system just below the grain (Fig. 2) and use forceps to separate the main roots, working from the top. Place these end to end along a ruler to measure their total length. Ignore the lateral roots for this purpose (unless you have plenty of time and patience).

(m) When you have made the measurements, place the roots and leaves in the container labelled '$+$' so that their dry weight can be found later.

(n) Repeat the measurements for each of the seedlings in turn, placing the leaves and roots in the appropriate container afterwards.

(o) Plot histograms (Fig. 3) of the root and shoot lengths for each seedling.

Culture solution	$+$	Ca	N	$-$
Abnormality of leaf colour				
Total leaf length				
Total root length				
Dry weight (whole class)				

Experiment 10.   Discussion

1   Which solution of mineral salts provided (a) the best and (b) the worst medium for the growth of the seedlings as judged by leaf length?

2   From your knowledge of plant nutrition, explain why nitrates (source of nitrogen) and calcium should be so important to a green plant.

3   Why would it be difficult to judge whether the lack of a particular element did more harm to the root growth than the shoot growth or vice versa?

4   Why do you think a small-seeded plant, rather than a runner bean, was used for this experiment?

5   Why do you think the solutions were topped up with distilled water rather than with the appropriate culture solution?

6   In what ways is this experiment unrepresentative of natural conditions?

7   Why was a culture solution lacking carbon not included, bearing in mind that plants need carbon for making their carbohydrates?

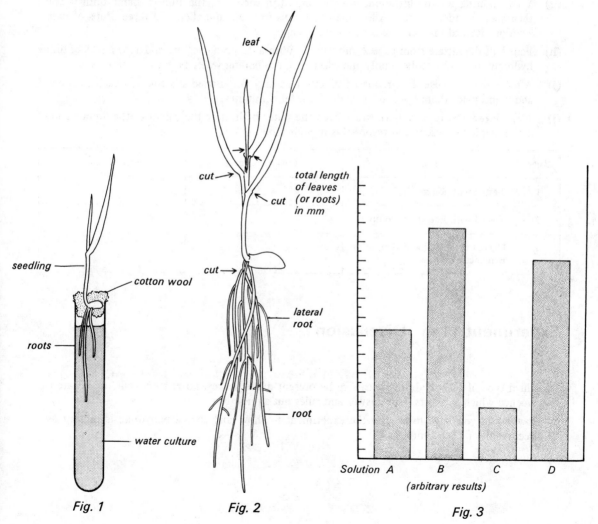

Fig. 1   *Fig. 2*   *Fig. 3*

Experiment 11a.   Carbohydrate production in iris leaves

(a) Label three test-tubes 1-3.

(b) Prepare a water bath by half filling a beaker or tin can with water and heating it on a tripod over a Bunsen burner. While waiting for the water to boil, copy the table below into your notebook. When the water boils, turn down the flame so that the water is kept just at boiling point.

(c) Dip the iris leaf in the boiling water for a second using forceps or by holding each end in turn.

(d) Hold the leaf over a mortar and cut it into small pieces with scissors. Add about 2 g (half a teaspoon) clean sand to the mortar and grind leaf and sand until the leaf is powdered.

(e) Use a graduated pipette to add 15 cm^3 water to the mortar and grind the mixture for a few seconds more.

(f) Pour about 5 cm^3 (20 mm in a test-tube) of the suspension into test-tube 1 and pour the rest into a fluted filter paper in a filter funnel inserted in test-tube 2.

(g) While waiting for the filtration, heat the liquid in tube 1 in the Bunsen flame until it boils, then cool it under the tap. Allow the cell debris to settle and then add three drops of iodine solution. Record in your table any colour change.

(h) Pour half the filtrate from tube 2 into tube 3. With a dropping pipette add TWO DROPS of dilute hydrochloric acid to tube 3 only and place it in the boiling water bath for 5 minutes.

(i) After 5 minutes, use the graduated pipette to add 5 cm^3 Benedict's solution each to tubes 2 and 3 and place them both in the water bath for 5 minutes.

(j) After 5 minutes, remove both tubes from the water bath, note the colour of the liquids and of any precipitate present and record this in your table.

Tube	Test	Colour change	Interpretation
1	Tested with iodine		
2	Tested with Benedict's solution		
3	Heated with HCl and then with Benedict's solution		

Experiment 11a.   Discussion

1 Which type of carbohydrate appears to be present in iris leaves, apart from cellulose? State the evidence which supports your answer and rules out alternatives.

2 How would you have to modify the experiment to show that the carbohydrate is present as a direct result of photosynthesis?

Experiment 11b.   The need for carbon dioxide; using iris leaves

The iris leaves have been in darkness for at least 4 days. If time permits, test one leaf for sugar as described below from **(f)** and record the result. If time is short, omit this test and assume that the leaves are relatively free from sugar.

(a) Label two screw-top jars A and B and add your initials. In A place about 20 g soda-lime. In B place about 20 cm^3 saturated sodium hydrogencarbonate solution (a test-tube full).

(b) In each jar place a specimen tube containing about 10 cm^3 water (50 mm in a test-tube).

(c) Insert one iris leaf in each tube of water, cut off any part of the leaf protruding from the jar and screw the lids on securely.

(d) Place the jars in front of a source of illumination or in a position where they will receive daylight but not direct sunlight. Copy the table below into your notebook.

(e) Leave the jars for 3-4 days in the light and then test each leaf as follows:

(f) In a beaker or can heat some water till it boils and, using forceps, dip the leaf from jar A into the boiling water for a second. Cut the leaf into small pieces over a mortar, add about 2 g clean sand and grind until the leaf is finely crushed.

(g) Use a graduated pipette to add 10 cm^3 water to the mortar and grind the mixture for a few more seconds before filtering it through a fluted filter paper into a test-tube labelled 'A'. While waiting for filtration, wash the mortar and start preparing leaf B in the same way, from **(f)**.

(h) When about 20 mm filtrate has collected in the tube add TWO DROPS only of hydrochloric acid and place the tube in the boiling water bath for 5 minutes.

(i) When tube A has been in the water bath for 5 minutes, use the graduated pipette to add 5 cm^3 Benedict's solution and return the tube to the boiling water for another 5 minutes. Meanwhile, continue the preparation and test with leaf B.

(j) Record in your table the colour in each tube after 5 minutes heating with Benedict's solution.

	Result of Benedict's test
Test on leaf after 4 days in darkness	
A. No carbon dioxide present (soda-lime)	
B. Carbon dioxide present (sodium hydrogencarbonate)	

Experiment 11b.   Discussion

1 From the colour of the liquid and precipitate in the test-tubes, which leaf contained more sugar?

2 What evidence is there that sugar is produced during photosynthesis?

3 How can you account, in terms of photosynthesis, for the fact that one leaf produced more sugar?

4 After as much as 4 days in darkness, sugar can still be detected in many iris leaves, while it takes only 48 hours darkness for starch to disappear from the leaves of those plants which normally produce starch. How can you account for this difference?

Experiment 12. Chromatographic analysis of chlorophyll

NOTE. The solvents used in this experiment are very inflammable. All Bunsen burners and other flames should be extinguished.

(a)   Pour about 10 cm³ solvent into the jar so that the bottom is covered to a depth of about 3 mm. Replace the lid and continue with instruction **(b)**.

(b)   With a pair of scissors, cut up about 1 g freshly collected grass, put the cuttings into a mortar and add about 2 g washed sand.

(c)   With a pestle, grind the sand and grass until it forms a pulp.

(d)   Add about 5 cm³ acetone (about 20 mm in a test-tube) and continue grinding for a further half minute.

(e)   Remove the pestle and allow the mixture to settle for a few seconds.

(f)   Take a fine glass pipette, tilt the mortar to separate some of the clear green solution from the grass pulp and dip the end of the pipette in this liquid. The green solution will run into the pipette.

READ ALL of instruction **(g)** before continuing.

(g)   Very briefly apply the tip of the pipette to a piece of chromatography paper about 15 mm from one end. The chlorophyll solution will run on to the paper and spread out. Try to produce a spot about 5 mm in diameter.

(h)   Allow the solution to dry for a few seconds and apply the pipette once again to the centre of the spot so that another drop of solution spreads out to the edge of the first spot. Repeat this ten times altogether, reloading the pipette from the mortar as necessary, and allowing the spot to dry between each application, so obtaining a dark green spot of chlorophyll.
If the mortar contents dry up before you have finished, add a few more drops of acetone.

(i)   Press on one side of the lid, thus slightly opening the slit,. and feed the chromatography paper through the slit till it just touches the solvent, with the chlorophyll spot about 10 mm above the liquid. DO NOT MOVE THE JAR.

(j)   Watch what happens to the solvent and the chlorophyll for a few minutes and then leave the jar, away from direct sunlight, for between 10 and 20 minutes, until the solvent has travelled up the paper nearly to the lid.

(k)   When the solvent is nearly at the lid, remove the lid and carefully pull the paper down through the slot and allow it to dry.

(l)   When the paper is dry, use a pencil or ball-point pen to **(i)** mark the outline of the colours and **(ii)** describe the colours (because they will fade in time). Hold the chromatography paper against the window to confirm the location of any faint areas of colour.

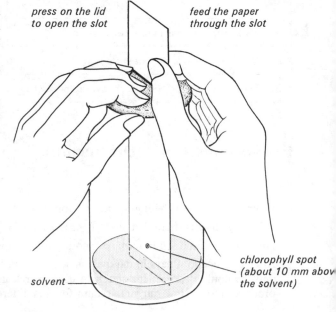

press on the lid to open the slot

feed the paper through the slot

solvent

chlorophyll spot (about 10 mm abov the solvent)

5 Germination and Tropisms

Introduction

Experiments 1 to 3 are concerned with the conditions for germination; experiments 6 to 11 test the responses of roots and shoots to the directional stimuli of light and gravity, i.e. tropisms. Experiments 10a and 10b are alternative methods of studying phototropism; 10b is suitable if a clinostat is not available but the two experiments examine slightly different aspects of the same phenomenon and it would be quite appropriate to attempt both. Experiments 12a and 12b are alternatives and there is little point in trying both; 12a is easier to prepare and conduct but gives rather less information.

Many of the experiments need seedlings which have been germinating for several days and before embarking on a series of experiments it is advisable to draw up a timetable for soaking and planting the seeds. Two weeks is the longest germination period needed.

The fruits of sunflower and maize may be difficult to obtain from shops at certain times of the year and it is best to buy these by weight, rather than by the packet, from seed merchants or biological supply houses.

Apart from the problem of obtaining seeds, there is no reason why the experiments should not be conducted at any time of the year. Most of the final trials for this section were carried out between November and February.

Contents

Experiment 1.   Is oxygen needed for germination?

(a)   Label two conical flasks A and B, and add your initials.

(b)   Roll a piece of cotton wool into a ball about 20 mm in diameter and tie it to a thread. Tie the other end of the thread on to the hook provided in the rubber bung and place the bung in the mouth of one flask. If necessary, adjust the length of thread so that the cotton wool hangs well clear of the bottom of the flask (Fig. 2).

(c)   Repeat this operation with another piece of cotton wool for the second flask.

(d)   Remove the bung from each flask, moisten the cotton wool and roll it in a dish of cress seeds so that the seeds stick to the wet cotton wool (Fig. 1).

(e)   In flask A place about 1 g pyrogallic acid and to this add about 10 cm³ (50 mm in a test-tube) sodium hydroxide solution.

CARE.   Any sodium hydroxide spilled on the skin or clothing must be washed away at once with much water. Sodium hydroxide spilled on the bench must be neutralized with dilute hydrochloric acid and wiped up.

(f)   In flask B pour 10 cm³ sodium hydroxide solution.

(g)   Securely replace both bungs with the cotton wool and seeds attached; place both flasks where they will be at the same temperature and degree of illumination and leave them for a week.

(h)   After one week, examine both flasks to see if the cress seeds have germinated and record the results in your notebook in a table like the one below.

	Extent of germination after 1 week
Flask A Sodium hydroxide & pyrogallic acid	
Flask B Sodium hydroxide only	

NOTE.   Sodium hydroxide and pyrogallic acid (sodium pyrogallate) absorb oxygen and carbon dioxide from the atmosphere. Sodium hydroxide absorbs only carbon dioxide.

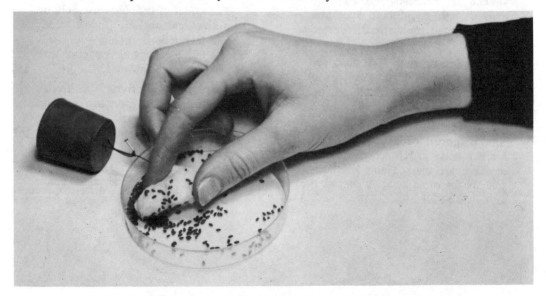

Fig. 1.   Roll the moist cotton wool in a dish containing cress seeds

Experiment 1. Discussion

1 What gases were lacking in flask A and flask B?

2 What difference was there in the germination of the seeds in the two flasks?

3 What seems the most likely explanation for this difference?

4 Do you have enough evidence to apply this explanation to all kinds of seeds?

5 What experiments would you have to conduct to be able to give a confident answer to question 4?

6 Why was it desirable to leave both flasks in the same conditions of temperature and illumination?

7 What was the point of setting up the experiment in flask B as well as in flask A?

8 Why was sodium hydroxide, rather than just water, placed in flask B?

9 Suppose it is argued that the seeds in flask A had been *killed* by the lack of oxygen or the presence of chemicals, how could you check on this?

10 Apart from the explanations offered in your answer to question 3 and in question 9, what other explanations could there be for the seeds' failure to germinate? In each case, suggest an experiment (not necessarily restricted to a school laboratory) which would help to eliminate the alternatives.

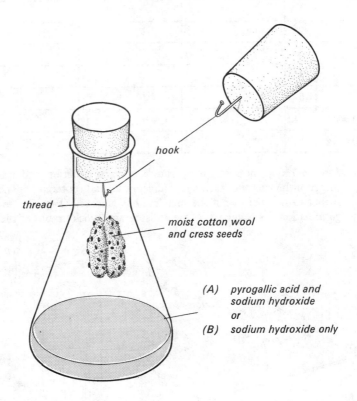

hook

thread

moist cotton wool
and cress seeds

(A) pyrogallic acid and
sodium hydroxide

or

(B) sodium hydroxide only

Fig. 2

Experiment 2.   Does temperature affect germination?

You are provided with three strips of blotting paper, three plastic bags and some soaked maize fruits.

(a)   Moisten one of the strips of blotting paper at the sink and let the surplus water drain away.

(b)   Place 5 maize fruits about 20 mm from the top edge of the moist blotting paper and about 50 mm apart. (Fig. 1)

(c)   Check that the grains are the 'correct' way up (Fig. 2) and roll up the blotting paper to trap the fruits between the layers.

(d)   Place the roll in a plastic bag. Write the letter A and your initials on a piece of paper and put this in the bag too.

(e)   Wrap the plastic bag round the paper roll and, leaving a space of at least 100 mm above the roll, fold down the top edge of the bag and secure it with paper clips (Fig. 3).

(f)   Place the bag with the paper roll upright in a container inside an oven or incubator at 25-30 °C.

(g)   Repeat the operation with the other two strips of blotting paper labelling them B and C respectively. Place B in a cupboard or under a box in the laboratory and C in a refrigerator at about 4 °C. Leave all three for 3-7 days.

(h)   In your notebook draw up three tables like the one below.

Temp °C		Length of shoot in mm	Length of main root in mm	Number of lateral and adventitious roots	Length of leaf (if present)
	1				
	2				
	3				
	4				
	5				
Total					
Average					

(i)   After 3-7 days note the temperature of each situation and unroll each blotting paper roll. Measure the length of each shoot (coleoptile) and root and note also the number of adventitious roots and whether or not the first leaf has emerged (Fig. 4). Do not record fruits in which no germination has occurred. Make a histogram (block graph) of the results.

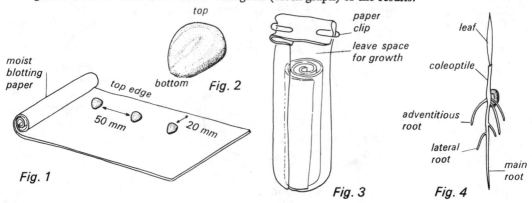

Fig. 1   Fig. 2   Fig. 3   Fig. 4

Experiment 2. Discussion

1 What effect has a rise in temperature on the rate of germination in maize?

2 Is this effect likely to be true of all seeds? Explain your answer.

3 Did a rise in temperature affect root and shoot growth to the same extent? Quote figures to support your answer.

4 What would you expect to happen if you tried the experiment at 40 °C and at 50 °C?

5 Why does your experiment *not* enable you to say what is the best temperature for the germination of maize?

6 How could you find out if the low temperature in the refrigerator had killed the fruits rather than slowed down their germination?

7 Why was it necessary to exclude light from experiment B?

Experiment 3. Is water essential for germination?

(a) Label three containers with your initials and the letter A, B or C. Put some dry cotton wool in each one.

(b) Place an equal number (about 10) soaked seeds on the cotton wool in each container.

(c) Leave A quite dry; add water to B until the cotton wool is uniformly moist; add water to C until the seeds are completely covered with water. Make sure the seeds do not float.

(d) Replace the lids on all three containers and leave them for a week in a situation where they will all be at about the same temperature.

(e) Draw up a table in your notebook, like the one below.

		Number of seeds germinating
A	Dry	
B	Moist	
C	Covered with water	

(f) After a week, remove the lids, count the number of seeds which have germinated in each container and record this number in your table.

Experiment 3. Discussion

1 In which containers, if any, had the seeds failed to germinate?

2 What aspect of the conditions in the container do you think was responsible for any failure to germinate?

Experiment 4.   The role of cotyledons in germination

(a)   Remove the testa from the beans as follows:
    (i)   hold the seed in the left hand with the hilum to the left;
    (ii)   run the thumbnail of the other hand down the right-hand edge of the seed from top to bottom, so removing a strip of testa (Fig. 2);
    (iii)   peel the testa off to the left, holding the cotyledons firmly together.

(b)   Carefully prise the cotyledons apart from the side opposite to the hilum and separate them completely so that the embryo is attached to one of them (Fig. 3). Discard the other cotyledon. Repeat **(a)** and **(b)** with the other beans until you have four cotyledons with embryos attached.

(c)   Leave one of the cotyledons intact and treat the other three as follows;
    (i)   Place the cotyledon on a Petri dish lid or tile and cut away $\frac{3}{4}$ of the cotyledon (Fig. 1).
    (ii)   Place another cotyledon on the dish or tile and cut away all the cotyledon except for a small area beneath the plumule.
    (iii)   Push a pin into the embryo on the third cotyledon at the point where the discarded cotyledon was attached to it and, keeping the embryo on the pin, prise it free from the cotyledon (Fig. 4).

(d)   Pin each of the portions of bean seed to the polystyrene block with the embryos towards the blotting paper and the radicles pointing downwards. When the pin is securely in the block withdraw it about 1 mm to leave the embryo just clear of the blotting paper (Fig. 5).

(e)   Moisten the blotting paper and put the block into a screw-top jar, add about 10 mm water and replace the lid. Use a spirit marker to write your name and the date on the outside of the jar and leave the jar in moderate light (not direct sunlight) for a week.

(f)   Copy the table below into your notebook.

(g)   After a week, remove the polystyrene block and the seedlings and make the measurements indicated in the table (see Fig. 6).

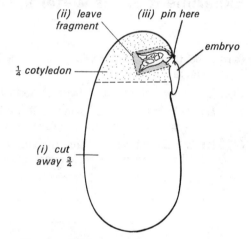

Fig. 1.   *Embryo with varying amounts of cotyledon attached*

	1 cotyledon	$\frac{1}{4}$ cotyledon	Cotyledon fragment	No cotyledon
Length of radicle				
Length of epicotyl				
Number of lateral roots				
Total growth 'score'				

(h)   Add up the measurements for each seedling to obtain a 'score' for the degree of growth in each case and make a block graph to show the differences.

Experiment 4.   Discussion

1 State what part you think the cotyledons play in the germination of a bean seed, justifying your statement with evidence from the experimental results.

2 In some cases there may not be much difference between the results with the whole cotyledon and the quarter cotyledon, or the results may appear to be 'the wrong way round'. What aspect of the experimental design might explain an unexpected result such as this?

3 Outline further experiments which could be carried out (not necessarily in a school laboratory) to support your statements in answer to question 1 and say what results you would expect.

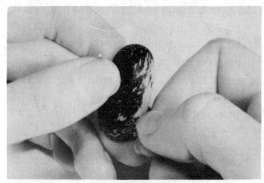

Fig. 2.   Remove the testa

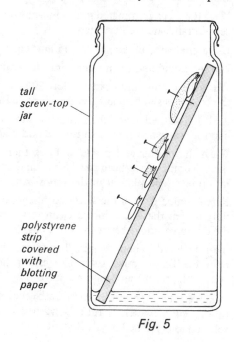

Fig. 5

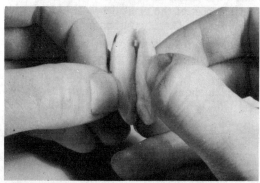

Fig. 3.   Separate the cotyledons

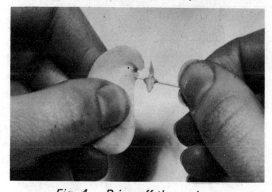

Fig. 4.   Prise off the embryo

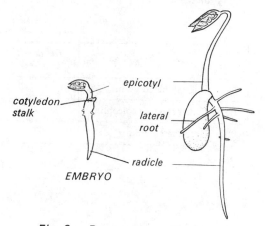

Fig. 6.   Runner bean after 1 week

Experiment 5. The use of food reserves in germinating cereals

You are provided with 20 ungerminated cereal grains which have been soaked for 12 hours and 20 cereal seedlings which have been germinating for 5-7 days.

(a) Prepare a water bath by half filling a beaker or can with water and heating it on a tripod and gauze over a Bunsen burner. When the water boils turn the flame down so that the water is just kept boiling. While waiting for the water to boil, copy the table below into your notebook.

(b) Label three test-tubes A-C.

(c) Place the 20 ungerminated grains in a mortar, add 3 g washed sand and grind with a pestle till the cereal grains are well crushed.

(d) Use a graduated pipette to add 15 cm³ tap-water and continue grinding for about half a minute.

(e) Tip the liquid and plant debris into test-tube A and place this in the water bath.

(f) Wash the mortar and pestle. Select 20 seedlings, cut off the roots and discard them. Cut off all the coleoptiles and shoots, letting them fall into the mortar.

(g) Add 3 g washed sand and grind the shoots to a pulp as before, finally adding 10 cm³ water. Tip the water and pulp into tube B and place this in the water bath.

(h) Wash the mortar and pestle and use them to grind up the 20 remaining grains from which the coleoptiles have been cut, using sand and 10 cm³ water as before. Tip the liquid and pulp into tube C and place it in the water bath.

(i) Remove tube A from the water bath, cool it under the tap and allow the debris to settle. Carefully decant the liquid into a clean test-tube leaving most of the plant debris behind. Divide the decanted liquid between two tubes, one of which is tube A duly washed out.

(j) To tube A add about 20 mm Benedict's solution and return it to the water bath for about 5 minutes. To the other tube of decanted liquid, add 4 or 5 drops of iodine solution and, in your table, record any colour change.

(k) Remove tube B from the water bath, cool it, decant the liquid, divide the liquid between two tubes, one of which is tube B washed out. Test the two tubes with iodine and Benedict's solution as described in (j). Record the results in your table. (Tube A should be ready by now.)

(l) Remove tube C from the water bath and test its contents as described in (i) and (j). Record your results.

		Result with iodine	Interpretation	Result with Benedict's	Interpretation
A	Ungerminated grain				
B	Cereal shoot & coleoptile				
C	Germinated grain				

Experiment 5. Discussion

1 In the ungerminated grain, what carbohydrates were revealed by your tests?

2 What carbohydrates were shown to be present in the germinating seedlings as a whole?

3 **(a)** Name any carbohydrate which was present in the germinating seedling but was not present in the ungerminated grain.

    (b) From what source could this additional carbohydrate have come

        (i) if the seedling were grown in light

        (ii) if the seedling were grown in darkness?

4 In the germinating seedling, describe any difference in the carbohydrate content of the shoot and the grain.

5 Suggest a sequence of events in a barley grain, germinating in the dark, which could account for the distribution of the food materials you have found.

6 If the seedling had been germinated in the light, what alternative explanation might be given for the change in type and distribution of carbohydrates.

7 Suggest experiments which would help to decide between these two explanations.

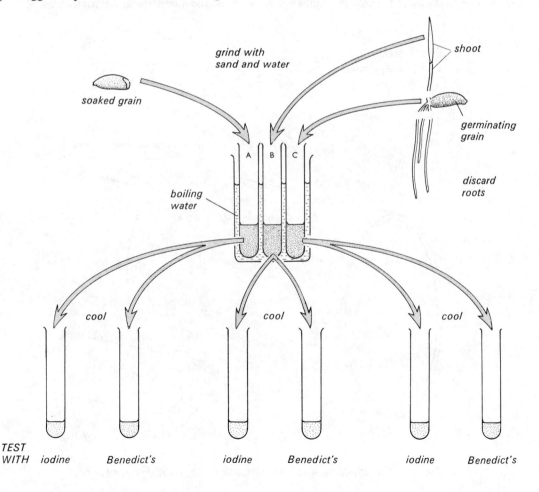

141

Experiment 6.   The response of radicles to unilateral gravity

You are provided with two pea seedlings with fairly straight radicles.

(a)   Make an accurate drawing of each seedling, similar to Fig. 1, in your notebook, showing any kinks or bends in the radicle. Leave space for two more drawings.

(b)   Measure each radicle in millimetres and write the length over the appropriate drawing.

(c)   By pushing a pin through the cotyledons, fix seedling B to one of the numbers marked on the pin board or cork inside box B, so that the radicle is horizontal (Fig. 3).

(d)   Pin seedling A to the same number on the turntable of the clinostat.   Make a note of your number.

(e)   Both sets of seedlings will be kept at a high level of humidity by putting wet blotting paper round the inside walls of the containers. Box B will remain stationary while the turntable A will be rotated slowly in a horizontal position (Fig. 2).

(f)   After a period of from 1 to 6 days, remove the lids and BEFORE unpinning your seedlings note any change in the direction in which the radicles are pointing.

(g)   Make an accurate drawing of each seedling in your notebook, alongside or below your previous drawings. Measure the radicle lengths again and write them over the appropriate drawing.

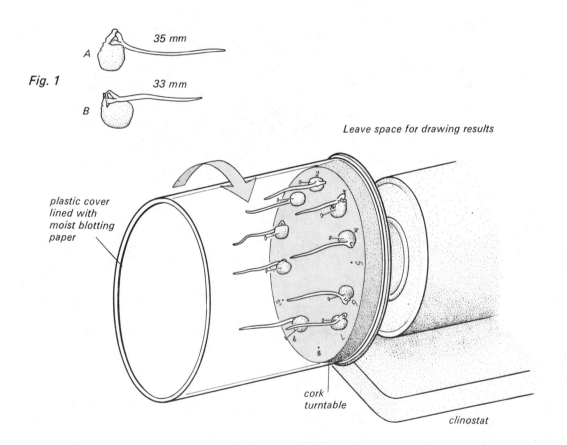

Fig. 1

A   *35 mm*

B   *33 mm*

Leave space for drawing results

plastic cover lined with moist blotting paper

cork turntable

clinostat

Fig. 2.   Pin seedling A to the turntable with the radicle horizontal

Experiment 6.   Discussion

1   Had either of the radicles grown? If so, by how much?

2   Had either of the radicles changed its direction of growth from the horizontal position to a recognizably new direction?

3   In what direction *with respect to the radicles* would gravity have been acting in (a) the stationary box, (b) the rotating box?*

4   If there was a change in direction of growth, was it related in any way to the direction in which gravity was acting? If so, what was the relationship?

5   If there was a recognizable change in the direction of growth of either radicle, was this change confined to the new growth since the experiment was set up or had it taken place in the part of the radicle which you drew on the first day?

6   If a pliable rod, e.g. a strip of plasticine, is held horizontally it might bend downwards under its own weight. What evidence have you that if a radicle appears to bend downwards it is not just bending under its own weight?

7   Assuming that your results, or the class results as a whole, suggest a relationship between the direction of gravity on a root and its direction of growth, what value would this have in the natural course of germination of a seed?

8   Do all the roots and branch roots of a plant's rooting system respond in the same way to unidirectional gravity?

9   Were your experiments adequately controlled, i.e. is there any other factor which might have influenced the direction of growth?

* 'Downwards' is not an acceptable answer. Gravity will act downwards on the seedlings whether they are horizontal or vertical, rotating or stationary.

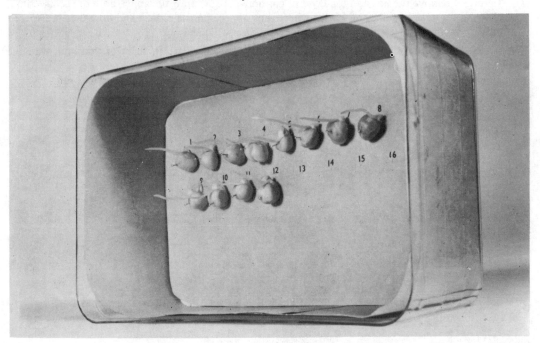

Fig. 3.   Pin seedling B in the stationary box

Experiment 7. The region of growth and response in radicles

You are provided with four pea seedlings with fairly straight radicles and a device of wire and cotton for marking them.

(a) Dip the cotton in the ink, wipe off any drops and use the inked cotton to make a series of marks along each of the radicles. Start from the very tip, spacing the marks out evenly and as close together as possible without letting them run together (Fig. 1).

(b) Place a strip of cotton wool in one side of a Petri dish as in Fig. 2 and run a little water on to it to make it moist but do not compress it.

(c) Arrange two of the marked seedlings as shown in Fig. 2 and cover them with a second piece of cotton wool leaving the radicles clear.

(d) Run water over the top layer of cotton wool to saturate it but do not let it run on to the radicles or it will wash off the marks.

(e) Replace the lid and hold it in position with an elastic band. Ensure that the lid holds the seedlings firmly in place and that the radicles are not touching the base or lid of the dish. Repeat this procedure for the second pair of seedlings.

(f) Write the letter A on one lid and B on the other, and put your initials on both. Stand each dish on edge so that the radicles in A are pointing vertically downwards and those in B are horizontal (Fig. 3). With a spirit marker put a line on the uppermost edge of the Petri dish (Fig. 4).

(g) Place the dishes, still on their sides with the mark upwards, in a box and leave them for 2 days.

(h) After 2 days, without removing the lids, examine the radicles through the Petri dish looking for evidence of growth and change in direction of growth.

(i) Carefully remove the lids and top layer of cotton wool and draw the appearance of each seedling showing the exact number of marks, where they appear on the radicle and how they are spaced out.

If possible, answer the discussion questions while the seedlings are in front of you and available for close study.

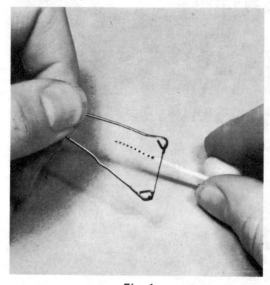

Fig. 1

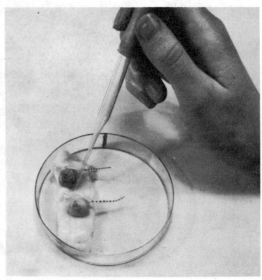

Fig. 2

Experiment 7.   Discussion

1   Were all the marks still equally spaced or were some spaced out more than others?

2   What conclusion would you reach about the growth of *a particular region* of the radicle if **(a)** the marks were still as close together as before, **(b)** the marks were more widely separated than before?

3   From the results with the seedlings in A, are you able to determine whether the radicle grows uniformly over its entire length or whether some parts grow more than others? Explain your answer.

4   From the results with the seedlings in B are you able to determine which part of a radicle responds to the stimulus of one-sided gravity? Explain your answer.

5   What was the reason for setting up dish A as well as dish B?

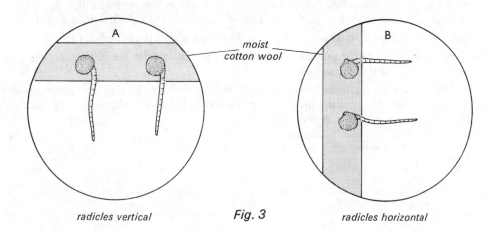

radicles vertical   *Fig. 3*   radicles horizontal

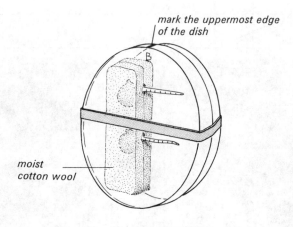

Fig. 4

Experiment 8.   The region of detection and response to one-sided gravity in roots

(a)   Label three Petri dishes A, B and C and add your initials to each.

(b)   Select 3 pea seedlings with straight radicles. Lie them alongside a ruler and, as accurately as possible, cut 3 mm from the root tip with a razor blade (Fig. 1).

(c)   Moisten 2 strips of cotton wool and wedge the seedlings between them in Petri dish A so that the seeds are held firmly between the base and the lid. The radicles must be horizontal and clear of the cotton wool and, as far as possible, not touching the sides of the dish (Fig. 3). Secure the lid with an elastic band.

(d)   Use a spirit marker to place dots on the lid and the base of the dish in line with the tip of each radicle (Figs. 2 and 3).

(e)   Select 3 more seedlings and carefully cut 1 mm from the tip of each radicle, as accurately as possible. Set up these seedlings in dish B in the same way as the others in dish A. Secure the lid with an elastic band and mark the position of the end of the radicle with dots as before.

(f)   Repeat the procedure with dish C and 3 more seedlings but leave the radicles intact.

(g)   Place all three dishes in the container provided so that the radicles are horizontal (Fig. 4) and copy the table given below into your notebook.

(h)   After 2 days, examine the three dishes WITHOUT OPENING THEM. Measure the increase in length of the radicle from the position indicated by the dot and note any change in the direction of growth. Record these results in your table.

	A 3 mm cut off		*B 1 mm cut off*		*C intact*	
	increase in length	*change of direction*	*increase in length*	*change of direction*	*increase in length*	*change of direction*
1						
2						
3						
Total increase						

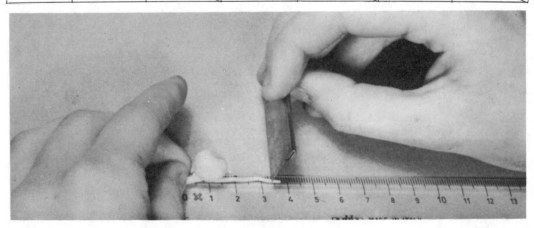

Fig. 1

146

Experiment 8. Discussion

1   Which, if any, of the radicles had changed their direction of growth and in which direction had they changed?

2   What effect does removal of **(a)** 3 mm, **(b)** 1 mm, from the tip of the radicle seem to have on the ability of a pea radicle to respond to one-sided gravity?

3   Which of the radicles, if any, had not significantly increased their length in 2 days?

4   What effect does removal of **(a)** 3 mm, **(b)** 1 mm, seem to have on the growth in length of the radicle?

5   From your results and those of the other members of your class, which parts of the pea radicle appear to be responsible for **(a)** continuing growth in length and **(b)** detecting or responding to the one-sided stimulus of gravity?

6   The result of experiment 7 suggests that the response to unidirectional gravity takes place in the region of the radicle which is growing most rapidly. From the results of experiments 7 and 8, what part does the final millimetre of root tip seem to play in **(a)** detection of, **(b)** response to, one-sided gravity?

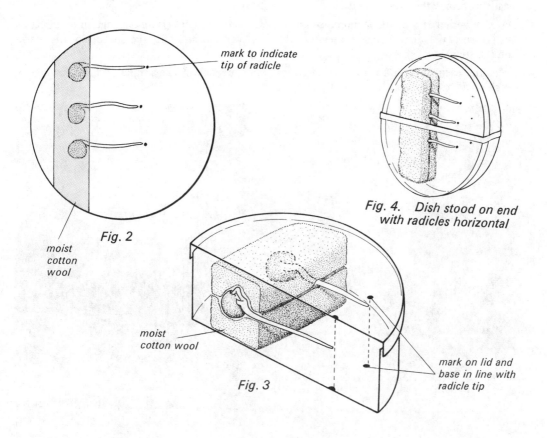

mark to indicate
tip of radicle

Fig. 2

moist
cotton
wool

Fig. 4.   Dish stood on end
with radicles horizontal

moist
cotton wool

Fig. 3

mark on lid and
base in line with
radicle tip

Experiment 9.　The response of shoots to unilateral gravity

You are provided with two bean seedlings with straight hypocotyls.

(a) Place the seedlings in turn alongside a ruler, measure the length of the shoot (Fig. 1) and record this length in your notebook.

(b) Use a spirit marker to put lines or dots on the hypocotyl and stem at 2 or 3 mm intervals, starting from the extreme tip. Count the number of marks on each stem and record this in your notebook.

(c) Fill two test-tubes with water. Wrap cotton wool round the base of the hypocotyl of each seedling and insert them in the mouths of the test-tubes with the root system fully immersed in water. The cotton wool should hold the seedling firmly in place and not let any water out when the tube is held horizontally. Mark each tube with your initials.

(d) Place one of the test-tubes on its side in the rack provided so that the shoot is horizontal and clear of the bench top (Fig. 2a).

(e) Place the other tube in one of the clips on the clinostat (Fig. 2b) so that the shoot is horizontal. The clinostat will be set to rotate with the seedlings held horizontally.　Leave both seedlings for 4-24 hours or more in a part of the laboratory which receives little or no direct light. Alternatively screen the experiments from strong light sources.

(f) After a period of 4 hours or more, examine each shoot and note **(i)** any change in its direction of growth, **(ii)** whereabouts in the stem this has occurred and **(iii)** the spacing of the marks on the stem.

(g) Measure each shoot and record its length in your notebook. Make simple outline drawings of each of the two shoots.

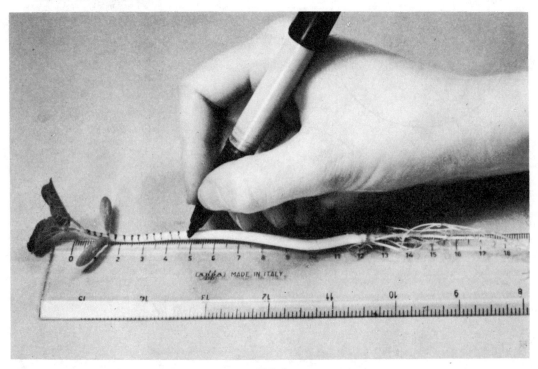

Fig. 1

Experiment 9. Discussion

1 Had the shoots increased in length? If so, by how much? Where, according to the marks on the stem, had most growth occurred?

2 Had the shoots changed their direction of growth, continued to grow horizontally, or had they not grown at all?

3 In what direction was gravity acting with respect to the shoots in (a) the stationary tubes, (b) the rotated seedlings*?

4 If either of the shoots had changed its direction of growth, in what way was this related to the direction of gravity?

5 If a change in direction had occurred, (a) was the curvature confined to the region which had appeared since you set up the experiment or (b) had it taken place in a region you had marked; (c) was it restricted to the region in which maximum growth had occurred?

6 Assuming that the class results as a whole suggest a relationship between the direction of gravity and the growth of a shoot, what advantage might this have in the normal germination of a seed?

7 Do all of the growing shoots of a flowering plant respond to gravity in the same way? Give one or two examples.

8 Why are the seedlings on the clinostat considered as controls in this experiment?

* 'Downwards' is not a satisfactory answer since it does not describe the direction in relation to the position of the plant, i.e. gravity will always act downwards whether the plant is upright or horizontal, stationary or rotating.

Fig. 2a. Rack   Fig. 2b. Clinostat

Experiment 10a. The effect of one-sided lighting on shoots

You are provided with two seedlings with straight shoots.

(a) Place the seedlings in turn alongside a ruler and use a spirit marker to put lines or dots at 2 mm intervals on the hypocotyl and stem, if any (Fig. 1). Count the number of marks on each shoot and record this in your book.

(b) Use the spirit marker to label two test-tubes A and B and add your initials.

(c) Fill both tubes with water. Wrap a strip of cotton wool round the base of the hypocotyl of each seedling and insert them in the mouths of the test-tubes with the root systems fully immersed in water (Fig. 2).

(d) Place tube A vertically in a container on the clinostat (Fig. 4) and tube B in the rack provided (Fig. 3).

(e) Arrange 60 W bench lamps 35-40 cm from both groups of seedlings and adjust their height so that the bulbs are level with a point 2-3 cm below the rim of the test-tubes (Fig. 2).

(f) Place sheets of cardboard round the experiments to screen them from direct light from windows, skylights or other sources and leave the plants for 1-3 days (Fig. 3).

READ ALL INSTRUCTION **(g)** before collecting your seedlings.

(g) After 1-3 days, remove the screens and *before* taking your own seedlings examine the whole group to see **(i)** whether the shoots have continued to grow vertically, **(ii)** whether they have altered their direction of growth but not in a consistent way or **(iii)** whether they have changed their direction of growth in a recognizable pattern.
If only some of the seedlings have changed their direction of growth note **(i)** how many and **(ii)** whether this direction is related to the side from which they received illumination.

(h) Take your own seedlings from the test-tubes and study the spacing of marks on the stem. Make a simple outline drawing of shoot B (the stationary seedling) indicating the direction from which the light was coming.

(i) Answer the discussion questions as far as possible while the seedlings are still available for study.

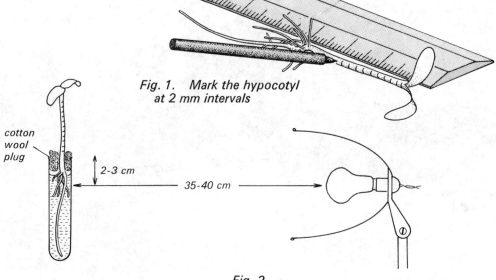

Fig. 1. Mark the hypocotyl at 2 mm intervals

cotton wool plug

2-3 cm

35-40 cm

Fig. 2

Experiment 10a.   Discussion

1 Considering the class results as a whole, what proportion of the shoots in **(a)** the clinostat, **(b)** the stationary rack had changed their direction of growth in the same way?

2 What relationship was there between any consistent change in direction of growth of the shoots and the direction of the light?

3 Judging by the spacing of the marks on the stem, had the shoot grown at all during the experiment? If so, where had the main growth occurred and in what way was the region of growth related to the region in which change of direction (if any) had taken place?

4 What effect would the rotation of the clinostat have on the direction of light reaching the shoots?

5 Assuming that the class results show a distinct relationship between the direction of light and the direction of growth of a shoot, what advantage might this relationship have when the seed germinates naturally?

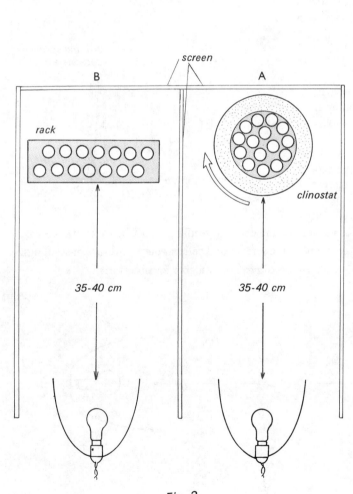

Fig. 3

Fig. 4.   Clinostat

Experiment 10b. The effect of one-sided lighting on cress seedlings

You are provided with two sets of cress seedlings growing on moist cotton wool.

(a) Use a spirit marker to label two Petri dish bases 1 and 2 and add your initials. Mark the base and sides of each dish as shown in Fig. 1.

(b) Use forceps to remove any seedlings which have fallen over or are leaning unduly.

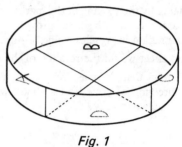

Fig. 1

(c) Use scissors to remove the cotyledons and extreme shoot tips of the seedlings growing on ONE HALF of the cotton wool (Fig. 2) in both sets.

(d) Place each group of seedlings in one of the marked dishes as shown in Fig. 3 and place dish 1 directly under a light source such as a bench lamp. about 20-40 cm away.

Place dish 2 so that it is receiving light from a source to one side and a few cm below the dish (Figs. 3 and 4).

(e) Rotate dish 2 so that sector A is towards the light source and 20-40 cm away from it if it is a bench lamp, or 5 cm away if it is a fluorescent tube.

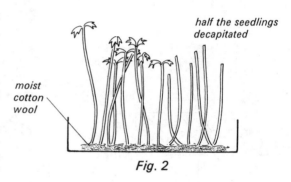

Fig. 2

(f) Thoroughly moisten the cotton wool and place cardboard screens round the experiment so that the seedlings cannot receive direct light from windows, skylights or other sources.

(g) If the experiment is to be left for more than one day, the seedlings must be watered each day.

(h) After a period of 90 minutes-7 days, remove the screens and collect your two dishes of seedlings.

(i) Answer discussion questions 1-3 while the seedlings are available for observation.

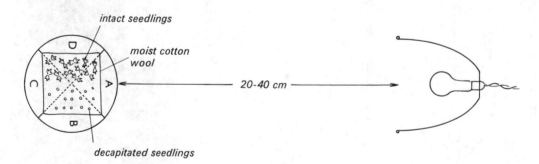

Fig. 3

Experiment 10b.   Discussion

1   Copy the table below into your notebook and examine both dishes to see in which direction the cress shoots are growing. Enter in the table the numbers of seedlings whose shoots are growing straight up and the numbers which are growing towards side A. Similarly count and record the numbers of shoots growing towards B, C or D.

growing towards	Light from side A		Light from above	
	intact shoots	*decapitated shoots*	*intact shoots*	*decapitated shoots*
A				
B				
C				
D				
straight up				

2   In dish 1 have the decapitated shoots grown as much as the intact shoots?

3   Consider those intact shoots in dish 2 which have changed their direction of growth. Has the curvature occurred in a region of the hypocotyl which is still present in the decapitated shoots?

4   From your results and answers so far, state what happens to cress shoots illuminated from one side and say what part you think is played by **(a)** the cotyledons and growing point and **(b)** the hypocotyl.

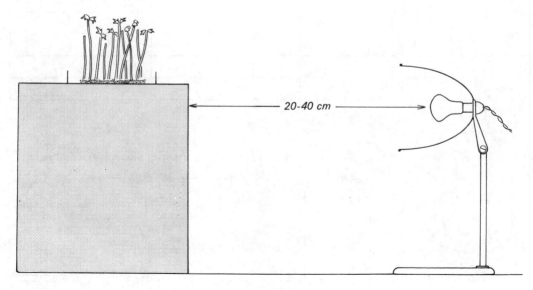

Fig. 4

Experiment 11. The region of detection and response to one-sided lighting in coleoptiles

(a) Copy the table given below into your notebook.

(b) Use forceps to remove any seedlings which are falling over, or with coleoptiles which are distorted, or less than about 10 mm, or more than about 50 mm.

(c)   (i) Use fine scissors to cut 5 mm from the coleoptile tips of one third of the seedlings.
  (ii) Use forceps to place aluminium foil caps over the tips of one third of the seedlings.
  (iii) Leave the remainder intact (Fig. 3).

(d) Measure the length of the coleoptiles in millimetres and record these figures in your table. It is not necessary to identify individual seedlings since only the totals will be compared.

(e) Use a spirit marker to write your initials on a Petri dish base and also mark a vertical line on the edge. Place the cotton wool and seedlings in the dish and moisten the cotton wool.

(f) Arrange the dish 15-30 cm from a bench lamp so that the mark on the dish is towards the lamp and the bulb is level with the edge of the dish (Figs. 1 and 2).

(g) Screen the experiment from direct light from other sources.

(h) After 3, 5 and 24 hours examine the seedlings and make drawings of a representative coleoptile from each group. After 24 hours, measure the coleoptiles and record the lengths in your table. If the first leaf has emerged, DO NOT include this in your measurement of the coleoptile. Calculate the total increase in length for each group of seedlings.

	Terminal 5 mm cut off			Terminal 5 mm covered			Intact seedlings		
	first length	second length	change in direction	first length	second length	change in direction	first length	second length	change in direction
Total									
Increase									

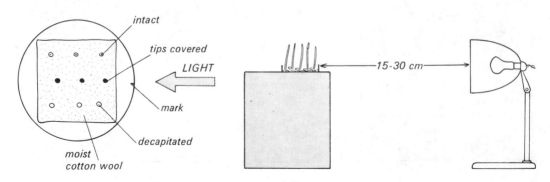

Fig. 1

Fig. 2

154

Experiment 11. Discussion

1 Which groups of coleoptiles had changed their direction of growth from vertically upwards and how was this related to the direction of the light source?

2 Consider the results with the seedlings whose coleoptile tips were covered. What effect did the exclusion of light from the terminal 5 mm have on (a) growth in length of the coleoptile, (b) response to one-sided light?

3 Did the foil cap exclude light from the region of the coleoptile which, in other seedlings, showed a growth curvature? In view of your answer, what role might the terminal 5 mm of coleoptile be performing in the phototropic response?

4 What effect had the removal of 5 mm from the coleoptile tip on (a) growth in length, (b) response to one-sided light?

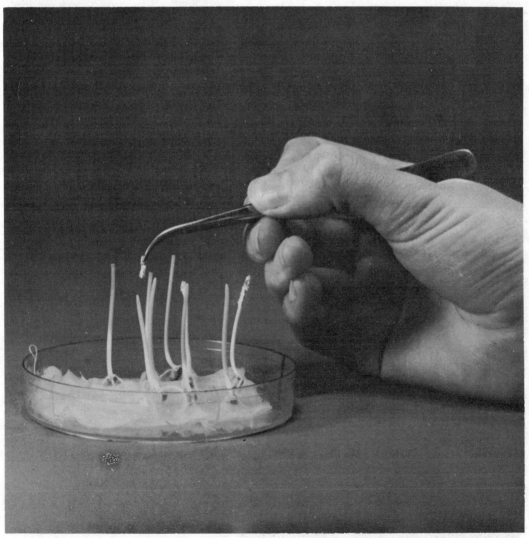

Fig. 3

Experiment 12a. The effect of indole acetic acid on coleoptiles

You are provided with two dishes containing moist cotton wool on which wheat or oat seedlings are growing.

(a) Copy the table on p. 157 into your notebook.

(b) Use forceps to remove from the cotton wool any grains which have **(i)** failed to produce a coleoptile of 10 mm or more, **(ii)** been tipped over by the growth of the root system, **(iii)** developed distorted or excessively long coleoptiles (over 40 mm), **(iv)** the first leaf emerging through the coleoptile.
In this way, reduce the number of seedlings in each dish to 10 even if it means removing healthy specimens.

(c) Label the dishes A and B with a spirit marker and add your initials to each.

(d) Measure the length of each coleoptile in millimetres by holding the cotton wool in one hand and letting it drape over the second finger so that the coleoptile can be held against a ruler (Fig. 1). Alternatively, use a pair of dividers. There is no need to identify individual seedlings since average lengths only are to be compared. Record the lengths in your table.

(e) Pick up a little lanolin on the end of a splint or match-stick and apply a blob of lanolin to the tip of each coleoptile in dish A. To do this, pick up the cotton wool and hold it as described in **(d)** so that the coleoptiles can be gripped between the thumb and first finger while applying the lanolin (Fig. 2a).

(f) Repeat the operation with seedlings from dish B but use a fresh splint and apply lanolin containing 1% indole acetic acid (IAA). Try to place the lanolin on the tip of the coleoptile and not on the sides.

(g) Moisten the cotton wool, place both dishes in the box provided and leave for 1-7 days (2 days is best).

(h) After 1-7 days take the cotton wool from each dish in turn, cut off the coleoptiles and measure them again, recording the lengths in your table. If the first leaf has emerged, do not measure this but only the coleoptile (Fig. 2b).

(i) Total the coleoptile lengths in each column and calculate the average length. Calculate the increase in length by subtracting the first average length from the final average length.

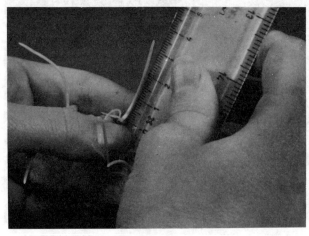

Fig. 1a

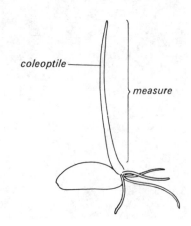

coleoptile —
} measure

Fig. 1b

Experiment 12a.   Discussion

NOTE.   The lanolin is used to keep a supply of IAA in contact with the tip of the coleoptile over a period of days.

1   Which group of treated seedlings showed the greater increase in length?

2   Do you consider that the difference in the two sets of results is large enough to be sure that it is not due to a chance variation between the two groups of seedlings?

3   Considering the class results as a whole, what would convince you that a difference in average coleoptile length between the two dishes could be attributed to IAA rather than chance?

4   What was the point of using plain lanolin on the seedlings in dish A if lanolin was merely a means of applying the IAA to the seedlings in dish B?

5   (a)   What effect does IAA appear to have on the growth of wheat or oat coleoptiles?
   (b)   In what ways might it achieve its effect?

6   What further evidence would you need to support the hypothesis that, in normal wheat or oat seedlings, the production of IAA in the coleoptiles promotes growth in length?

7   If IAA had been applied to the side rather than the tip of the coleoptile, what result would you have expected?

	A *lanolin only*		B *lanolin + indole acetic acid*	
	1st length	*2nd length*	*1st length*	*2nd length*
	10 Lines			
Total				
Average				
Increase				

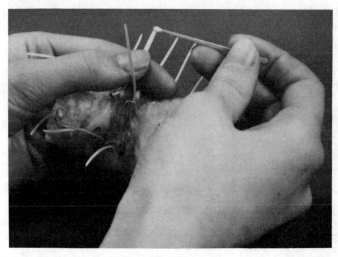

Fig. 2a

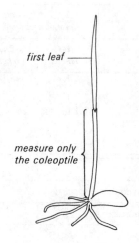

first leaf

measure only
the coleoptile

Fig. 2b

Experiment 12b. The effect of indole acetic acid on wheat coleoptiles

You are provided with four dishes containing moist cotton wool on which wheat seedlings are germinating.

(a) Copy the table given below into your notebook.

(b) Lift the dishes with the germinating seedlings from the box. Using forceps, remove from each dish any grains which have (i) failed to germinate, (ii) not produced a coleoptile about 10 mm long, (iii) been tipped over by the roots, (iv) developed distorted or excessively long coleoptiles (over 40 mm). In this way reduce the number of seedlings in each dish to 10, even if it means removing healthy specimens.

(c) With a spirit marker, label the dishes 1-4 and add your initials. In the case of 1, 2 and 3 (but NOT 4) cut 2 mm from the tip of each coleoptile using scissors or a razor blade.

(d) Measure the length of each coleoptile in millimetres by holding the cotton wool in one hand and letting it drape over a finger so that the coleoptile can be held against a ruler (Fig. 1). Alternatively use a pair of dividers. There is no need to identify individual seedlings since only the average length will be used. Record the lengths in the appropriate part of your table.

(e) Return dishes 3 and 4 to the box and moisten the cotton wool. Treat 1 and 2 as follows:
   (i) Pick up a little lanolin on the end of a splint or match-stick and apply a blob of lanolin to each coleoptile tip in dish 1 to cover the cut surface. To do this, pick up the cotton wool from the dish, drape it over the middle finger and grip the coleoptile with your first finger and thumb while applying the lanolin (Fig. 2).
   (ii) Do the same with 2 but use a fresh splint and lanolin containing 1% indole acetic acid (IAA). Make sure the cut end is completely covered.

(f) Return dishes 1 and 2 to the box, moisten the cotton wool and leave for 1-7 days (2 days is best).

(g) After 1-7 days, remove each of the dishes in turn from the box, cut off the coleoptiles and measure them again, recording the lengths in your table. If the first leaf has emerged, do not measure this. Measure only the coleoptile (see Fig. 2b, p. 157).

(h) Total the coleoptile lengths in each column and calculate the average length. Calculate the increase in length by subtracting the first average length from the final average length.

	1 *Tip cut and lanolin applied*		2 *Tip cut and IAA applied*		3 *Tip cut*		4 *Tip intact*	
	1st length	*2nd length*	*1st length*	*2nd length*	*1st length*	*2nd length*	*1st length*	*2nd length*
	10 Lines							
Total								
Average								
Increase								

Experiment 12b.   Discussion

NOTE.   The lanolin is used to keep a supply of IAA in contact with the cut end of the coleoptile.

1   Compare the increase in length of all four groups of coleoptiles. Which group showed **(a)** the greatest, **(b)** the least rate of growth?

2   What effect did the removal of 2 mm from the coleoptile tip appear to have on the growth rate of the coleoptile? What figures do you compare to make this judgement?

3   Did the application of plain lanolin to the cut coleoptile tip appear to increase the growth rate compared with the cut but untreated coleoptiles? If there was a significant difference, what explanation could there be?

4   Did the application of the lanolin containing IAA increase the growth rate compared with **(a)** the cut but untreated coleoptiles, **(b)** the cut coleoptiles treated with plain lanolin?

5   What was the point of using plain lanolin on the seedlings in dish 1 if lanolin was merely a means of applying IAA?

6   **(a)**   What effect does IAA appear to have on the growth of wheat coleoptiles?
    (b)   In what ways might it achieve this effect?

7   If wheat coleoptiles produced their own IAA, in what part of the coleoptile would you expect it to be produced? Give the evidence supporting your answer.

8   Do the results of your experiments offer any evidence to suggest that wheat coleoptiles do in fact produce IAA?

9   Suppose the coleoptiles do produce their own IAA, do your results suggest whether it would be at a concentration of more or less than 1%? Explain your answer.

10   If IAA in lanolin had been applied to only one side of the cut coleoptile, what result would you have expected?

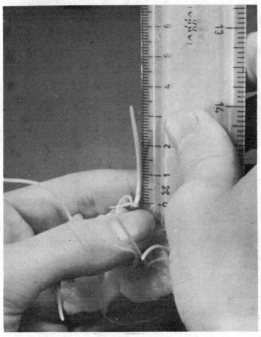
Fig. 1

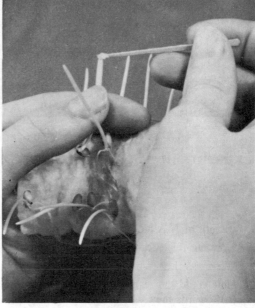
Fig. 2

Experiment 13. The effect of indole acetic acid on maize coleoptiles

You are provided with two maize seedlings with straight coleoptiles.

(a) Label two test-tubes A and B, add your initials and fill the tubes with water.

(b) Wrap each seedling in a strip of cotton wool and insert it in the test-tube so that the roots are immersed in water and the seedling is held upright and firm.

(c) Use a clean splint to collect a little lanolin and place a blob of it on one side of coleoptile A about 5 mm from the tip.

(d) Use a fresh splint to collect a little lanolin which contains 1% indole acetic acid (IAA) and place this on one side of coleoptile B about 5 mm from the tip.

(e) Return both test tubes to the rack and keep in darkness for 2 days or more.

(f) After 2 days examine each seedling and make drawings of the appearance of the coleoptiles and the position of the lanolin blobs.

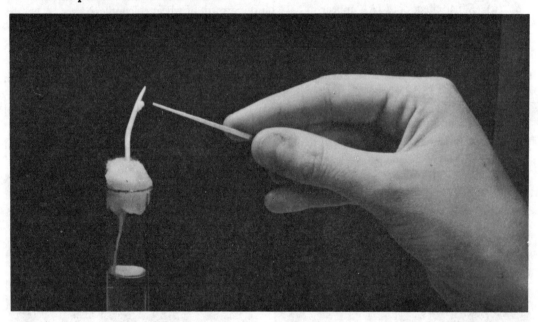

Experiment 13.   Discussion

1   From the results of experiment 12a or b (IAA on wheat coleoptiles) what effect do certain concentrations of IAA have on the growth of coleoptiles?

2   How can this effect be used to explain the appearance of coleoptile B?

3   What was the reason for treating coleoptile A with plain lanolin?

4   How could the results with coleoptile B be used to explain the growth curvature of coleoptiles (a) towards light, (b) away from gravity, assuming that coleoptile tips actually produce IAA or a substance having similar effects?

Experiment 14. Respiratory activity in maize seedlings

Respiration is the process by which energy from food is made available for chemical reactions in the cells of living organisms.

This process involves removing hydrogen atoms (oxidation) from the food and passing them to other chemicals known as hydrogen acceptors.

Tetrazolium chloride is an 'artificial' hydrogen acceptor which goes pink in solution when it has taken up hydrogen atoms.

You are provided with two soaked maize fruits one of which has been boiled. You are also provided with a 6-day-old maize seedling.

(a) Prepare a water bath by half filling a beaker or jar with water, adjusting the volumes of hot and cold water to give a temperature of about 40 °C.

(b) Label three test-tubes A, B and C. Use a graduated pipette to put 2 cm³ tetrazolium chloride solution into both A and B, and 6 cm³ into C. Place all three tubes in the water bath.

(c) Place the fruits and the seedling on a tile or Petri dish lid. Using a scalpel or razor blade cut the soaked fruits in half longitudinally so that the cut passes through the long axis of the embryo (Fig. 1). Drop both halves of the boiled grain into tube A and the half grains of the other fruit into tube B. Cut longitudinally through the mid line of the seedling in the same way, starting by splitting the coleoptile down its length. Continue the cut through the fruit and if possible down the length of the radicle. If the radicle is too long for this, cut it into 20 mm lengths and split these. Put all the seedling parts into tube C and leave all the tubes for 10 minutes.

(d) During the 10 minutes, copy the outlines of Figs. 2 and 3 into your notebook.

(e) If the cut surfaces of the fruits do not show much pink colour after 10 minutes leave them for another 10 minutes and check that the water has not cooled below 30 °C.

(f) Pour the solution back into the beaker and shake the fruits out on to the bench. Blot them dry, examine each cut surface with a hand lens and shade your drawings to show both the distribution and intensity of colour in B and C.

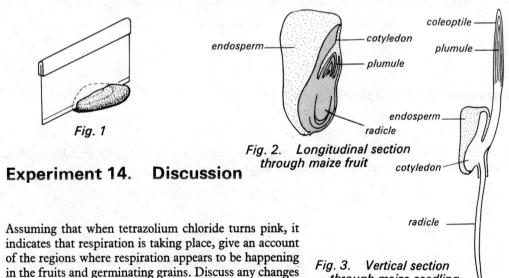

Fig. 1

Fig. 2. Longitudinal section through maize fruit

endosperm

cotyledon

plumule

radicle

coleoptile

plumule

endosperm

cotyledon

radicle

Fig. 3. Vertical section through maize seedling

Experiment 14. Discussion

Assuming that when tetrazolium chloride turns pink, it indicates that respiration is taking place, give an account of the regions where respiration appears to be happening in the fruits and germinating grains. Discuss any changes in the distribution of these areas during the course of germination.

Experiment 15.   The effect of light on shoots

(a)   Write your name and the date on two strips of blotting paper (double thickness). Moisten the blotting paper.

(b)   Select 10 pea seedlings at equivalent stages of germination. Roll 5 of them up in one of the blotting paper strips so that the radicles are pointing downwards and the epicotyls (shoots) are just clear of the top of the blotting paper (Fig. 1).

(c)   Repeat this operation with the other 5 seedlings, making sure that your initials are visible on the outside of the roll. Place each roll separately in one of the two plastic bags or containers provided and copy the table given below into your notebook.

(d)   One of the containers will be left in darkness for a week while the other will receive continuous illumination.

(e)   After a week, open the blotting paper rolls and make a brief report in your book, comparing the colour of leaves, size of leaves and colour of stem.

(f)   Make and record the measurements indicated in the table, starting with the seedlings grown in the light. Dissect off the leaves and stipules with forceps if necessary (see Fig. 2).

In light				In darkness			
length of shoot (a)	total no. of leaflets & stipules (b)	no. of internodes (c)	length of internodes (calculate from a & c)	length of shoot	total no. of leaflets & stipules	no. of internodes	length of internodes

Experiment 15.   Discussion

From your observations, state briefly what effect light seems to have on shoots and discuss those features of the seedlings which seem to be little affected by light.

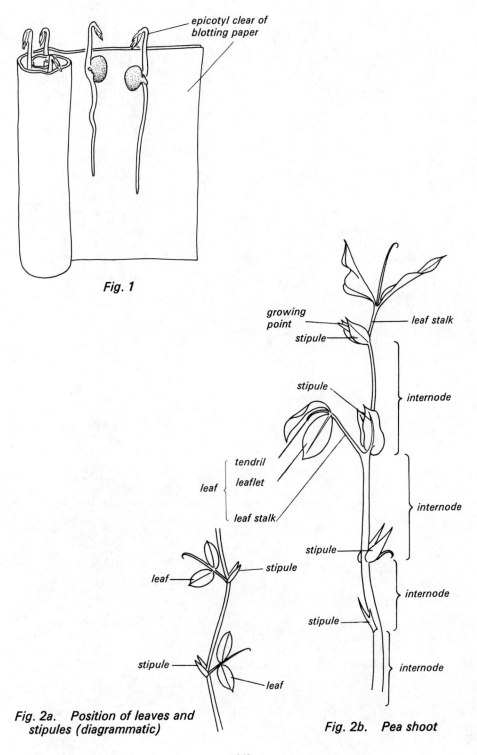

epicotyl clear of
blotting paper

Fig. 1

growing
point

stipule

leaf stalk

stipule

internode

tendril

leaflet

leaf

internode

leaf stalk

stipule

internode

stipule

leaf

stipule

stipule

internode

stipule

leaf

Fig. 2a.   Position of leaves and
stipules (diagrammatic)

Fig. 2b.   Pea shoot

163

6 Diffusion and Osmosis

Introduction

These experiments, if carried out in the order in which they are written, introduce a logical sequence of ideas, starting with simple diffusion in gases and liquids, extending them to the special case of water diffusing through membranes and finally observing some of the effects of osmosis on living material. None of the experiments is intended as an alternative to any other but it is not necessary to attempt all the experiments in order to elucidate the principles of osmosis and diffusion.

Most of the experiments can be attempted at any time of the year but experiments 9, 11, 12 and 15 use plant material which is not readily available in the winter months.

With the exception of experiments 2, 3 and 10, results should be obtained easily within an hour or less.

Contents

Experiment 1. Diffusion in gases

You are provided with two glass tubes equal in length and diameter and marked at 2 cm intervals.

(a) Copy the table given below into your notebook and collect a little cold water in a beaker or other container.

(b) Using forceps, pick up a square of red litmus paper, dip it in the water, shake off surplus water and place the paper inside one of the glass tubes. Use the wire or glass rod to manipulate the litmus paper until it is stuck to the glass immediately under the 10 cm mark (see Fig. 2). Repeat this operation for the remaining marks in both tubes, working from the end of the tube nearer the marks.

(c) Close the 28 cm end of both tubes with the ordinary cork bungs.

The next operation involves ammonia solutions, one of which is very strong and gives off a pungent vapour. It is harmless enough provided you do not deliberately sniff it at close range.

(d) Take the corks with cotton wool plugs to the central dispensing point in the laboratory and use the dropping pipette to place about 20 drops of strong (9N) ammonia solution on the cotton wool in cork A and an equal number of drops of dilute (2N) ammonia solution on the cotton wool in cork B.

(e) *Note the time* and insert each cork in the appropriate tube at the same time.

(f) In your table, note the time interval required for each square of litmus to turn completely blue and continue recording until the litmus at 28 cm in one of the tubes has turned blue.

(g) Make a graph of distance against time, plotting the values for both tubes on the same graph, with time on the horizontal scale.

Time	Distance along tube in cm	10	12	14	16	18	20	22	24	26	28
started	A Strong ammonia solution Number of minutes from start										
....	B Weak ammonia solution Number of minutes from start										

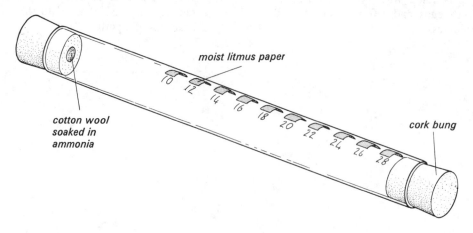

cotton wool soaked in ammonia

moist litmus paper

cork bung

Fig. 1

168

Experiment 1. Discussion

In order to turn the litmus blue, molecules of ammonia must have travelled from the cotton wool at one end of the tube to the litmus paper at the other end. This process of molecular movement is called DIFFUSION.

1 Why is it unlikely that (a) air currents through the tube or (b) convection currents inside the tube could have distributed the ammonia?

2 What influence on the rate of diffusion does the concentration of the source have in this case?

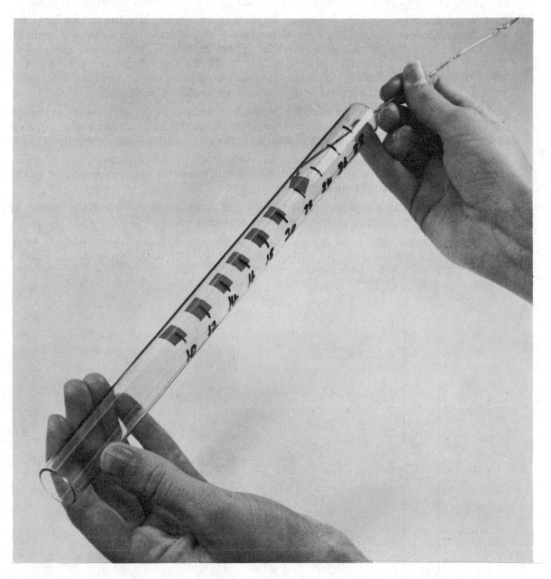

Fig. 2

Experiment 2.   Diffusion in liquid

You are provided with a boiling tube containing 10% gelatin.

(a) Melt the gelatin in the tube by immersing it almost up to the rim in a beaker or jar of hot water (**Fig. 3, p. 173**). **While waiting for it to melt continue with (b).**

(b) Use a spirit marker to label three test-tubes A, B and C and make three marks on tubes A and B as shown in Fig. 1. Add your initials to these two tubes.

(c) Pour the liquid gelatin into tubes A and B up to mark 1 and allow it to set. Setting can be hastened by dipping the tube upright into cold water. Do not allow it to set with the surface oblique.

(d) *Read the whole paragraph before proceeding.* Pour about 20 mm liquid gelatin into tube C and use a dropping pipette to add one drop of methylene blue solution. Swirl the tube gently to mix the dye with the gelatin. Make sure the gelatin in tube A has set and use a clean dropping pipette to transfer enough blue gelatin to fill the space between marks 1 and 2. Insert the pipette well down inside tube A so that blue gelatin does not touch the sides above mark 2 (see Fig. 3). Use the pipette to draw off any air bubbles which form.

(e) Add another 9 drops of methylene blue solution to tube C and use the dropping pipette to transfer this blue gelatin to tube B in the same way as before. Allow the blue gelatin in tubes A and B to set firmly (about 5 minutes in cold water). While waiting, continue with instruction **(g).**

(f) When the blue gelatin has set, run a little cold water into both tubes to ensure that no liquid gelatin remains. Pour off the water and fill both tubes up to mark 3 with cool but liquid gelatin and cool it quickly. Cork both tubes (Fig. 2).

(g) In your notebook make diagrams, similar to Fig. 1, of tubes A and B to show the position of the gelatin and the blue dye. Leave space for two more diagrams.

(h) After a week, examine the tubes again and make two more diagrams beside the first two to show the distribution and intensity of the blue colour. Answer discussion questions 1 and 4 while the tubes are still available for examination.

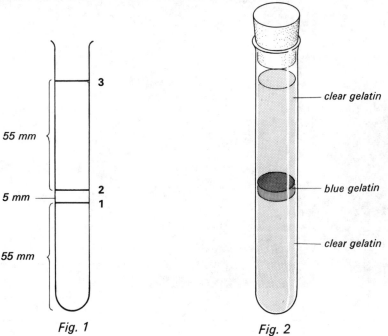

Fig. 1   Fig. 2

Experiment 2.   Discussion

1 Did the diffusion of methylene blue take place equally upwards and downwards in the tube? Give figures to support your answer.

2 In what direction would you expect diffusion to occur if a drop of methylene blue were surrounded on all sides by a large volume of gelatin?

3 If you did experiment 1, comment on the relative speeds of diffusion in air and in gelatin.

4 What difference was there in the rates of diffusion of methylene blue in tubes A and B, bearing in mind that the concentration in tube B was about ten times greater than in tube A? Give measurements to support your answer.

NOTE.   You may not regard gelatin as a true liquid but water would have been unsatisfactory for two main reasons: (a) it is difficult to start with a distinct boundary between two liquids which can mix and (b) convection currents can occur and so distribute the dye. The gelatin prevents the water from flowing so that the dye can move only by diffusion.

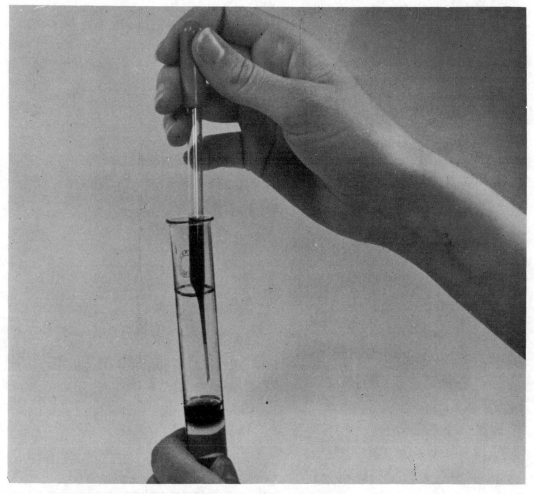

Fig. 3. Do not let the blue gelatin touch the sides of the tube

Experiment 3. Two-way diffusion in a liquid

(a) Melt the gelatin in the boiling tube by immersing it almost up to the rim in a beaker or jar of hot water (Fig. 3). While waiting for the gelatin to melt, continue with **(b)**.

(b) Use a spirit marker to label two test-tubes A and B and draw lines round them at 8, 12 and 13.5 cm from the bottom of the tube (see Fig. 1).

(c) Pour liquid gelatin into both tubes up to the *first* mark and use a dropping pipette to add 10 drops of 1% cresol red (a pH indicator) to each. Close each tube in turn with your thumb or a rubber bung and invert it several times to mix the indicator with the gelatin.

(d) Allow the coloured gelatin to set firm. This can be hastened by dipping the tubes in cold water. Meanwhile, make three copies of Fig. 1 in your notebook, labelling only the first one.

(e) When the yellow gelatin has set, remelt the gelatin in the boiling tube if necessary and pour a further layer of cool but liquid gelatin into both tubes up to the second mark and cool quickly.

The next operation uses ammonia solution which gives off a pungent gas. It is harmless enough provided you do not sniff it at close range.

(f) When the clear gelatin has set in tube A, add 1% ammonia solution up to the third mark and insert the bung (Fig. 2). Do the same for tube B but use 10% ammonia solution.

(g) Put your initials on the tubes and leave them on their sides for 3-7 days.

(h) After 3-7 days, examine the tubes and use crayons or shading to indicate the distribution and colour of the indicator on your two remaining drawings.

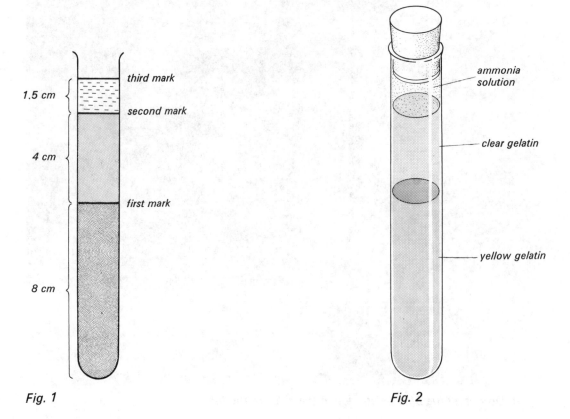

Fig. 1

Fig. 2

Experiment 3.   Discussion

NOTE.   Assume that diffusion alone is responsible for the movements of ammonia and cresol red.
  Cresol red is a pH indicator, yellow in acid* and red in alkali.   Ammonia is an alkali.

1 What evidence have you that the ammonia has diffused through the gelatin?
2 What evidence is there to show that the cresol red has also diffused?
3 Write a sentence to compare the direction and distance of diffusion in the two substances.
4 How could you explain the fact that ammonia diffuses further in tube B than in tube A?
5 Does the diffusion of ammonia in one direction appear to affect the diffusion of cresol red in the other direction? What evidence is there to support your answer?

* (between pH 1.8–7.2).

Fig. 3. Dip the boiling tube in hot water to melt the gelatin

Experiment 4.   Diffusion and size

The gelatin block contains cresol red, a pH indicator which is red in alkali and yellow in acid.

(a) Copy the table given below into your notebook.

(b) Place the gelatin block on a tile or Petri dish lid and use a scalpel or razor blade to cut it in half (Fig. 1), producing two cubes of 10 mm side (1 cm cube).

(c) Keep one of these cubes intact and cut the other in half.

(d) Repeat this cutting operation as shown in the flow chart opposite so that you end up with five blocks of the dimensions shown in the table.

(e) Fill a test-tube to within 10 mm of the top with dilute hydrochloric acid.

(f) *Note the time* and, starting with the largest block, drop all the blocks into the acid in the test-tube and close it securely with a rubber bung or cork.

(g) Tilt the tube about to spread the gelatin blocks along its length. Hold the tube horizontally and rotate it so that you can see each block clearly and from all sides in turn. Try not to warm the tube too much with your hands or the gelatin may dissolve.
Note the time taken for the acid to penetrate to the centre of the block as indicated by the disappearance of the red colour. Record this time in your table.

	1	2	3	4	5
Cube dimensions in mm	10 x 10 x 10	10 x 10 x 5	10 x 5 x 5	5 x 5 x 5	5 x 5 x 2½
Time for acid to penetrate					

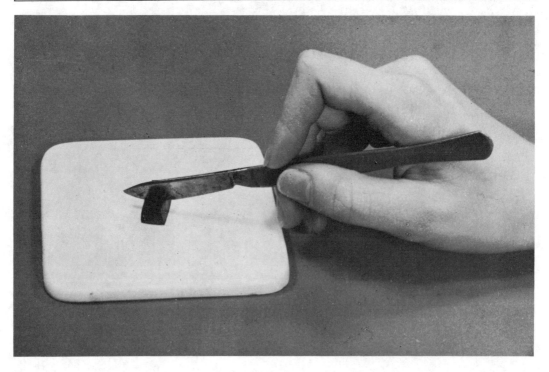

Fig. 1

Experiment 4. Discussion

1 In general terms, what is the relationship between the rate of penetration of acid into the gelatin block and the size of the block?

2 How much more rapidly did acid penetrate to the centre of the smallest block than to the largest block?

3 **(a)** Was the rate of penetration the same or nearly the same for any of the blocks?
(b) Suggest a reason to support your observation.

4 Imagine that, on a much smaller scale, the gelatin block represents a cell of a single-celled creature: **(a)** what substances need to penetrate to all parts of a living cell; **(b)** what substances need to diffuse out of the cell in the opposite direction?

5 Most single-celled animals are less than 1 mm across and the largest are only about 2 mm long. By reference to your results in this experiment suggest one reason why single-celled animals are no larger than this?

6 On the basis of the results of this experiment, the penetration of substances into a large, multi-cellular organism such as a fish should take a very long time. What aspects of structure and organisation in a fish allow it to survive and be active despite such a slow rate of diffusion?

7 Suggest one or more shapes for a volume of one cubic centimetre of gelatin, which would increase the rate of penetration of substances.

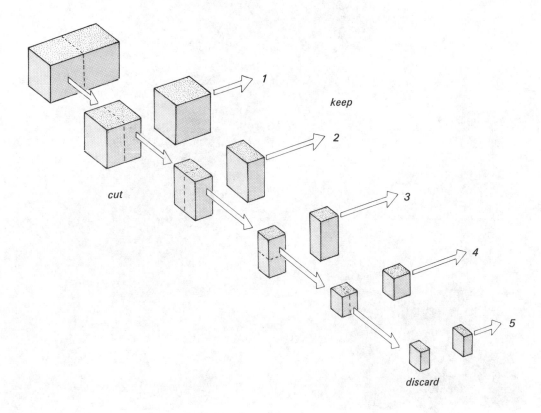

Fig. 2

Experiment 5. Differential permeability

You are provided with a length of dialysis tubing which has been soaked in water to make it flexible.

(a) If the dialysis tubing is not already knotted at one end, tie a knot close to one end of the tubing, leaving the other end open.

(b) Use a dropping pipette to half fill the dialysis tubing with starch solution (Fig. 1).

(c) Place the dialysis tubing with its starch solution inside a test-tube, fold the open end of the tubing over the rim of the test-tube and secure it with an elastic band as shown in Fig. 2.

(d) Wash the test-tube and dialysis tubing under a running tap to remove all traces of starch solution from the outside of the dialysis tubing and then fill the test-tube with dilute iodine solution and leave it in the rack for 10-15 minutes.

(e) After 10-15 minutes, examine the iodine and starch solutions in the test-tube and record any colour changes.

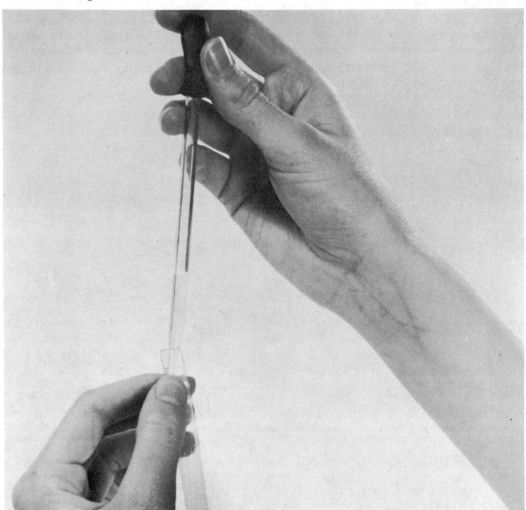

Fig. 1. Half fill the dialysis tubing with starch solution

Experiment 5.   Discussion

1 What colour change is usually seen when iodine solution is added to starch, or starch to iodine?

2 After 10-15 minutes, what colour was (a) the starch solution in the dialysis tubing and (b) the iodine solution in the test-tube?

3 How can you explain the fact that the same colour change had not occurred in both solutions? (There are two possible explanations.)

4 What result would you expect if the iodine solution was in the dialysis tubing and the starch solution in the test-tube?

5 What properties must the dialysis tubing possess if these results are to be explained?

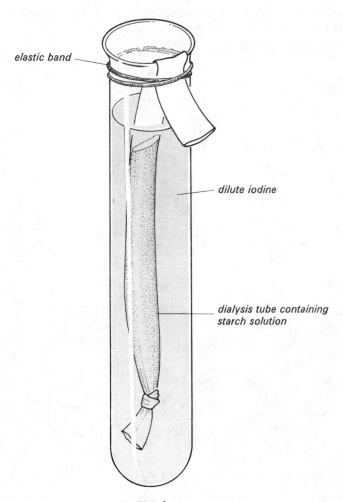

Fig. 1

Experiment 6. Osmosis

(a) Use a dropping pipette to fill the dialysis tube and connector with syrup solution (Fig. 1).

(b) Push the capillary tube into the connector far enough to bring the syrup to a level somewhere in the lower third of the capillary (Fig. 2).

(c) If there are short columns of liquid and air bubbles trapped in the upper part of the capillary, squeeze the dialysis tubing to force the syrup to the top of the capillary and then let it return slowly.

(d) Clamp the capillary tube vertically in a stand and then lower it to immerse the dialysis tubing in a beaker or jar of water until it is completely covered but not touching the bottom of the vessel (Fig. 3). Leave for a minute so that the syrup can acquire the temperature of the water.

(e) Move the rubber band on the capillary to mark the level of the liquid.

(f) Watch the level of the liquid in the capillary. If it falls rapidly, examine the dialysis tube through the side of the beaker to see if syrup is escaping.

Measure any change in level after 10 minutes.

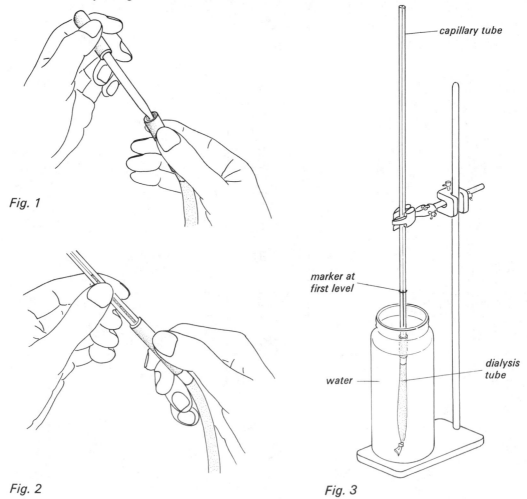

Fig. 1

Fig. 2

Fig. 3

capillary tube

marker at first level

water

dialysis tube

Experiment 6. Discussion

1 Assuming that there were no leaks, what happened to the level of liquid in the capillary?

2 What does this result indicate about the volume of liquid in the dialysis tubing and capillary?

3 What explanation can you offer for this change in volume?

4 Supposing that the capillary tube could be extended indefinitely and the experiment left for a long time, at what point would the experiment stop working, assuming that the water level in the beaker is maintained?

5 This experiment has been conducted without a control. What control do you think should be carried out to exclude any alternative explanations of the results?

6 In what ways is the result of this experiment similar to the result of experiment 5?

Experiment 7.   Dialysis

(a) Label four test-tubes A to D.

(b) Use a graduated pipette to place 3 cm³ starch solution and 3 cm³ glucose solution in tube A. Close the mouth of the tube with your thumb and invert it several times to mix the solutions.

(c) Securely knot one end of a length of dialysis tubing if this has not been done already and use the pipette to place 3 cm³ of the starch glucose mixture in it (Fig. 2).

(d) Place the dialysis tubing in tube B, fold the open end over the rim and secure it with an elastic band (Fig. 3).

(e) Wash away all traces of starch and glucose from the outside of the dialysis tube by filling and emptying tube B with water several times. Finally, fill the tube with water, note the time and leave the tube in the rack for 15 minutes or more.

(f) During the 15 minutes, copy the table given below into your notebook and prepare a water bath by half filling a beaker or can with water and heating it to boiling point on a tripod and gauze with a Bunsen burner. When the water boils, reduce the Bunsen flame to keep it just at boiling point.

(g) After 15 minutes, pour some of the water from tube B into tube C to a depth of about 20 mm, add an approximately equal volume of Benedict's solution and place tube C in the water bath for 5 minutes (Fig. 1).

(h) Pour another 20 mm water from tube B into tube D and add 5 drops of iodine solution.

(i) Record in your table the results of the iodine and Benedict's test.

Liquid from tube B tested with	Colour	Interpretation
Benedict's solution		
Iodine solution		

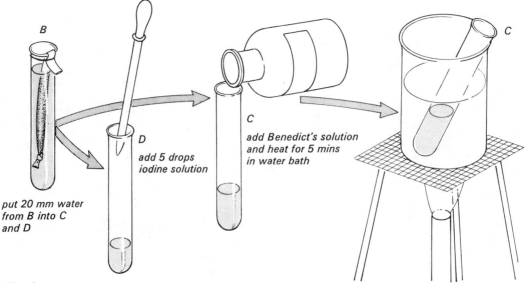

put 20 mm water from B into C and D

add 5 drops iodine solution

add Benedict's solution and heat for 5 mins in water bath

Fig. 1

Experiment 7.   Discussion

1 Which of the two substances in the dialysis tube was able to escape into the surrounding liquid and which was retained?

2 Can you think of any differences in physical or chemical properties of the two substances in solution which might account for the observed effects?

3 Blood plasma is a solution of mainly salts, glucose, proteins and urea. In the artificial kidney, the patient's blood is circulated through the dialysis tubing. If the tubing were immersed in water and dialysis occurred, which of the substances listed above would you expect to escape into the water?

4 How could the escape of the useful substances be prevented without altering the properties of the dialysis tubing?

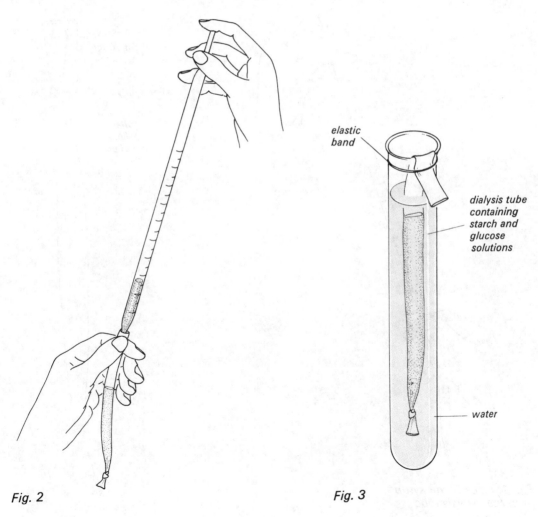

Fig. 2   Fig. 3

Experiment 8.   Turgor

You are provided with a length of dialysis tubing which has been soaked in water and knotted at one end.

(a) Use a graduated pipette to place 3 cm³ syrup solution in the dialysis tubing (Fig. 1).

(b) Taking care to exclude air bubbles from the tubing, tightly knot the open end of the dialysis tubing to leave only about 20 mm tubing free above the knot (Fig. 2).

(c) Wash the tubing under the cold tap to remove all traces of syrup solution from the outside. The partly filled tube should be flexible enough to bend through 90° (Fig. 3).

(d) Place the dialysis tubing in a test-tube and fill the test-tube with water (Fig. 4).

(e) Label the test-tube with your initials and leave it in the rack for 30-45 minutes.

(f) After 30-45 minutes remove the dialysis tubing filled with syrup and note how it differs from its starting condition.

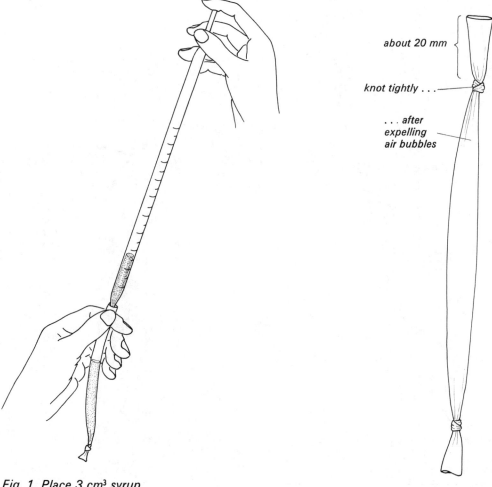

Fig. 1. Place 3 cm³ syrup in the dialysis tube

about 20 mm

knot tightly . . .

. . . after expelling air bubbles

Fig. 2

Experiment 8.   Discussion

1 What difference did you notice between the starting and final condition of the dialysis tube filled with syrup solution?

2 What change in (a) volume, (b) pressure, must have taken place in the syrup solution?

3 Suggest an explanation for the change which took place.

4 What might have happened if the experiment had been left for more than 45 minutes?

5 Suppose that, at the beginning of the experiment, the dialysis tubing had been inserted into a test-tube which it just fitted in diameter, what effect would this have on the result assuming water was available at the top and bottom of the test-tube?

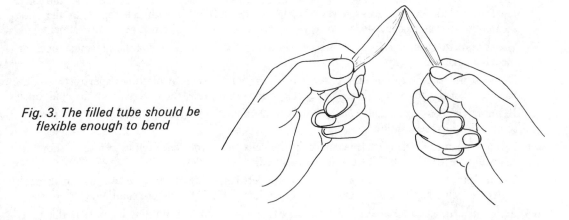

Fig. 3. The filled tube should be flexible enough to bend

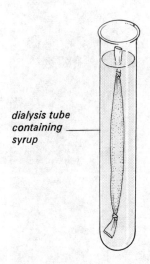

dialysis tube containing ⎯⎯ syrup

Fig. 4. Fill the test tube with water

Experiment 9.   Plasmolysis

(a) Use a scalpel or razor blade to make a shallow transverse cut in the red epidermis of the piece of rhubarb stalk.

(b) With a pair of fine forceps lift up a strip of the epidermis at one side of the cut. Lift only the epidermis and not the underlying cortex. Having freed a narrow band of epidermis, pull it off with the forceps (Fig. 2) and press it flat on a slide with the outermost surface upwards.

(c) Use the scalpel or razor blade to cut about 10 mm of this strip from the thinnest and reddest portion (Fig. 3) and, using a dropping pipette, cover this with 3 drops of water.

(d) Use the forceps to lower a cover slip carefully on to the water drops, avoiding trapping air bubbles (Fig. 4), and examine the epidermis under the microscope using the ×10 objective.

(e) Move the slide about to find a group of clearly defined cells *near the edge*, with red cell sap and make a drawing in your notebook to show one of these cells. Draw the cell at least 50 mm long, representing the outline accurately and shading the area filled with cell sap. Clip the slide securely to the microscope stage and leave it in this position for the rest of the experiment.

(f) Use the pipette to place 2 drops of sucrose solution on the left-hand side of the slide, just touching the edge of the cover slip.

(g) Draw all this solution under the cover slip by applying a strip of blotting paper to the right-hand edge of the cover slip (Fig. 1). Try not to move the slide, the cover slip or the epidermis.

(h) Examine the cells again and watch for about 2 minutes. If nothing happens, draw through some more sucrose solution.

(i) When a significant change has occurred in the cells, draw the same cell as before to show the cell wall and the cell sap. The cell is plasmolysed.

(j) Use the pipette to place 3 drops of water on the left hand side of the slide and draw it through under the cover slip as before. Do this twice to flush out all the sucrose solution.

(k) Study the cells again for about 2 minutes repeating operation **(j)** if nothing happens in this time.

Fig. 1. Draw the solution under the cover slip with blotting paper

Experiment 9.   Discussion

1 When the rhubarb cells were exposed to sucrose solution what change did you observe in **(a)** the shape of the vacuole and **(b)** the colour of the cell sap?

2 What change, if any, took place in the shape of the cell?

3 Bearing in mind the fact that liquids cannot be compressed, what must have happened to the cell sap to account for **(a)** the change in volume and **(b)** the change in colour?

4 After exposure to the sucrose solution, what do you suppose occupied the space between the vacuole and the cell wall in the plasmolysed cells?

5 Why did the cell sap not mix with the liquid in this space?

6 Which part of the cell must be 'semi-permeable' in order to explain these results?

7 How do the results of this experiment lead to the conclusion that the cell wall is permeable not only to water but also to dissolved sucrose?

8 What effect would it have on the tissues of the whole plant if all the cells were plasmolysed?

9 **(a)** What changes took place in the cells when the sucrose was replaced by water?
   (b) How can you explain these changes in terms of osmosis?

10 How would this change, if it applied to all the cells, affect the tissues of the plant as a whole?

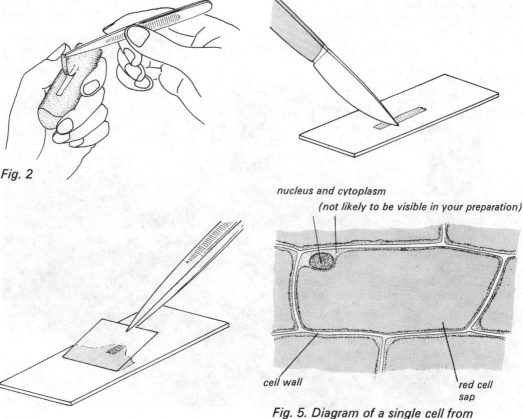

Fig. 2

Fig. 4

nucleus and cytoplasm
(not likely to be visible in your preparation)

cell wall

red cell sap

Fig. 5. Diagram of a single cell from the epidermis

Experiment 10.   Turgor in potato tissue

(a) Label three test-tubes A, B and C and add your initials.

(b) Use a graduated pipette to put 20 cm³ water in tube A and 10 cm³ in tube B.

(c) Similarly, put 20 cm³ 17% sucrose solution in tube C and 10 cm³ in tube B.

(d) Working on a dissecting board or folded newspaper, cut the ends off a large potato and use a No. 4 or 5 cork borer to obtain 4 or 5 cylinders of potato tissue as long as possible (Fig. 2).

(e) Push the cylinders of potato out of the cork borer with the flat end of a pencil (Fig. 3) and select the three longest.

(f) Cut all three to the same length, e.g. 50, 60 or 70 mm, trimming the ends at 90° at the same time (Fig. 1). Put one cylinder in each test-tube.

(g) Leave the potato tissue for 24 hours.

(h) After 24 hours, use forceps or a mounted needle to remove the potato cylinder from test-tube B. Rinse it in a beaker of water and measure its length in mm. Repeat this operation for the potato cores in tubes A and C noting at the same time whether the tissue is firm or flabby. Record your results in your notebook.

Initial length .... mm	*A Water*	*B 8.5% Sucrose*	*C 17% Sucrose*
Length after immersion in			

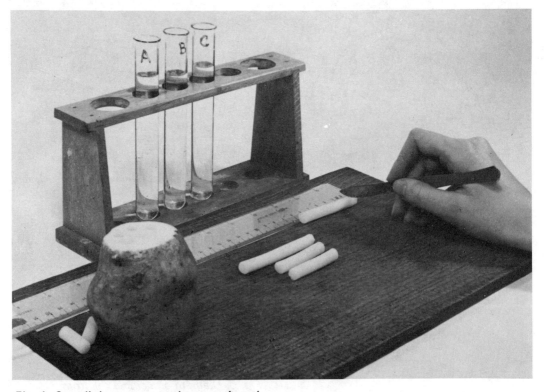

Fig. 1. Cut all three cores to the same length

Experiment 10.   Discussion

1 From your knowledge of osmosis in plant cells, what do you suppose has happened to the cells of the potato tissue (a) in 17% sucrose, (b) in water?

2 How would these changes in the cells account for any changes in size of the potato cylinder?

3 What do you think limits the change in length of the potato cylinder which was immersed in water?

4 How could the results of this experiment be incorporated in a hypothesis to explain the extension growth of plant shoots and roots?

5 How could the experiment be modified to find out the osmotic concentration of cell sap?

Fig. 2. Obtaining cylinders of potato tissue

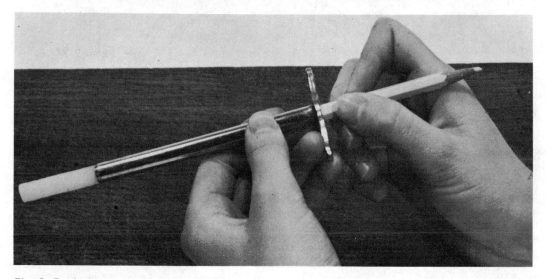

Fig. 3. Push the core out with a pencil

Experiment 11.   Turgor in a dandelion stalk

(a) Label two watch-glasses or Petri dishes A and B. In A place some tap-water and in B some salt solution.

(b) Use a razor blade or scalpel to cut a piece of dandelion stalk about 40 mm long and then split it lengthwise into strips about 1 or 2 mm wide (Fig. 1).

(c) Place one of these strips in the water in dish A and the other in the salt solution in dish B.

(d) Watch what happens to each strip in the next few seconds. If any curling takes place notice whether the epidermis side of the strip is on the inside or the outside of the curve.

(e) When no further change takes place, swap the strips over so that the one which was in the water is now in the salt solution and vice versa.

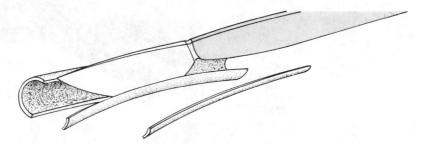

Fig. 1. Cut the stalk into strips (but not on the bench)

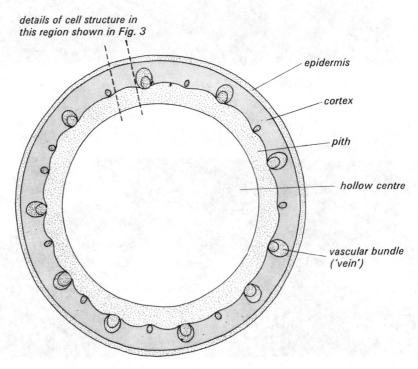

Fig. 2. Diagram of transverse section through dandelion stalk

Experiment 11. Discussion

1 What shape were the strips before they were placed in the dishes?

2 Describe what happened to the strips of stalk when they were placed **(a)** in the water and **(b)** in the salt solution, either at first or after you swapped them over.

3 Study Figs. 2 and 3, which show the microscopic structure of the dandelion stalk as seen in transverse section. From these diagrams and your knowledge of cell structure, offer an explanation for the behaviour of the strips of stalk when placed in **(a)** water and **(b)** salt solution, assuming that all the cell walls are equally permeable and all the cell sap has the same concentration.

4 How could the forces which produce the movement in dish A help to make the stalk firm and upright in the living plant?

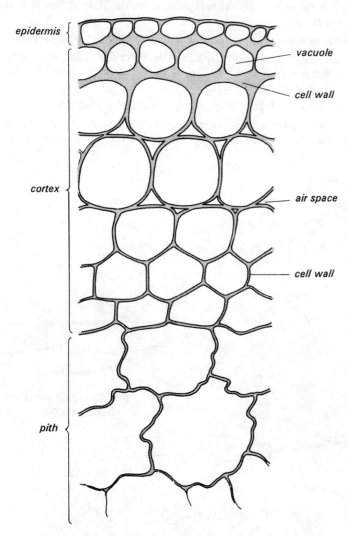

Fig. 3. Cell structure of dandelion stalk in transverse section
   (cell contents not shown)

Experiment 12. Stomatal movements

(a) Tear the leaf with a twisting and pulling action so that some of the lower epidermis is exposed (Fig. 1).

(b) Place a small piece of the leaf with an exposed fringe of epidermis on a microscope slide, lower surface uppermost, and trim off the thick part of the leaf with a scalpel (Fig. 2).

(c) Add 2 or 3 drops of water to cover the epidermis and carefully lower a cover slip over it, excluding all air bubbles (Fig. 3).

(d) Examine the slide under the microscope using the $\times$ 10 objective. Move the slide about until you find a group of stomata which are open (Fig. 4). Clip the slide firmly to the stage and leave it there for the rest of the experiment.

(e) With a dropping pipette, place 2 drops of sodium chloride solution on the slide so that it is in contact with one edge of the cover slip.

(f) Examine the stomata and be ready to observe them again immediately after the next operation.

(g) Place a square of blotting paper on the slide on the opposite side of the cover slip to the salt solution and move it so that it just touches the cover slip (see Fig. 1, p. 184). As soon as the blotting paper starts to draw the solution through, watch the stomata closely.

(h) If nothing seems to happen, draw through 2 more drops of sodium chloride solution.

(i) When the change is complete, try and reverse it by drawing through water instead of salt solution. The procedure may have to be repeated 2 or 3 times in order to wash away all the salt solution.

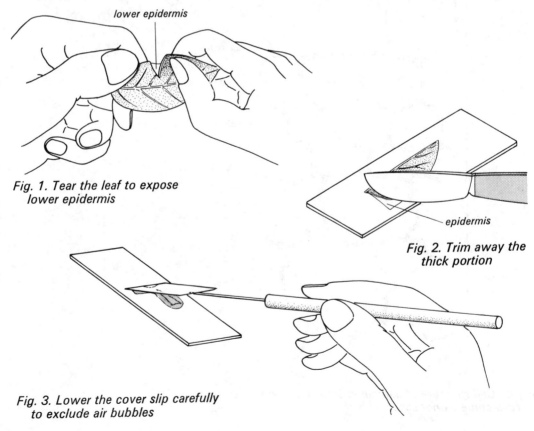

Fig. 1. Tear the leaf to expose lower epidermis

Fig. 2. Trim away the thick portion

Fig. 3. Lower the cover slip carefully to exclude air bubbles

Experiment 12.   Discussion

1 In what ways do guard cells differ in structure from the other cells of the epidermis?

2 What do you see happening to the stomata when the sodium chloride solution reaches the epidermis?

3 From your knowledge of osmosis, what would be the effect on the guard cells of a concentrated solution of sodium chloride? How would this effect influence the turgor of the cells?

4 Do the stomata close when the guard cells are turgid or when they are flaccid?

5 What natural events in the leaf of a plant could cause a change in turgor in the guard cells?

6 What change in shape of the guard cells must occur when the stoma closes?

7 Can you explain how a change in turgor could bring about such a change in shape? (Study Figs. 4-6.)

8 Stomata are usually open during part of the day but closed at night. Suggest a hypothetical mechanism connecting light, chemical reactions in the leaf, osmotic pressure, turgor, and structure of guard cells which would explain the opening and closing of stomata.

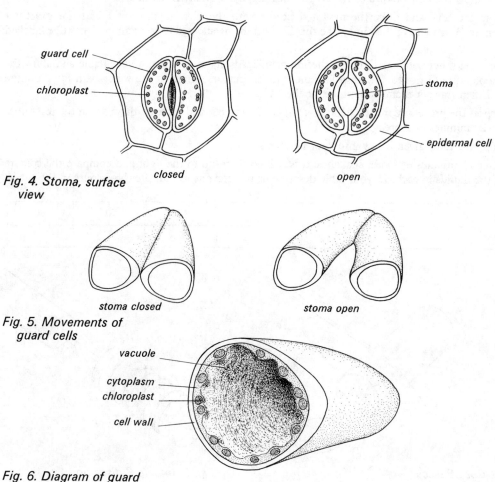

Fig. 4. Stoma, surface view

closed   open

Fig. 5. Movements of guard cells

stoma closed   stoma open

Fig. 6. Diagram of guard cell structure

Experiment 13: Selective permeability and temperature

(a) Use a cork borer to cut cylinders of tissue from a beetroot. Push the cores of tissue out of the borer with the flat end of a pencil. Place these cylinders of beetroot on a tile or Petri dish lid and use a scalpel or razor blade to cut them into discs about 3 mm thick. You will need 36 discs.

(b) Place the discs in a beaker or jar and wash them in cold water for at least 5 minutes. Either leave them under a running tap or fill and empty the container about five times during the washing process, but in either case continue with **(c)** and **(d)**.

(c) Label six test-tubes, each with one of the figures 30, 40, 50, 60, 70 and 80, and use a graduated pipette to put 6 cm³ of cold tap-water in each.

(d) Prepare a water bath by half filling a beaker or can with cold water, place it on a tripod and gauze and put a thermometer in the water but DO NOT start heating it yet.

(e) When the beetroot discs have been washed, impale 6 of them on a mounted needle with a space between each one (Fig. 2).

(f) Heat the water bath with a small Bunsen flame till the water reaches 25 °C, remove the burner and allow the temperature to rise to 30°, heating again briefly if necessary.

(g) Note the time and place the mounted needle with the discs in the water bath for exactly 1 minute. At the end of this time, take the discs off the needle and drop them into the tube labelled 30.

(h) Heat the water bath again till the thermometer reads about 37°, remove the flame and allow the temperature to rise to 40°. Impale 6 more discs, immerse them in the water bath for 1 minute and then transfer them to the tube labelled 40.

(i) Repeat the procedure for temperatures 50, 60, 70 and 80° and leave all the discs in the test-tubes for 20 minutes or more.

(j) Copy the table given below into your notebook.

(k) After 20 minutes or more, shake the tubes, hold them up to the light and compare the colours of the liquids in each. In your table describe or use crayons to indicate the colours in the tubes.

Temperature °C	30	40	50	60	70	80
Colour						

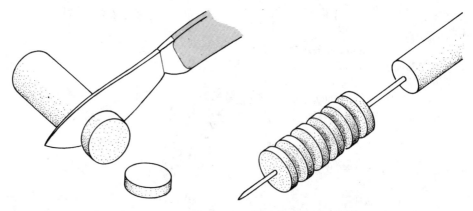

Fig. 1. Cut 36 discs
of beetroot

Fig. 2. Impale 6 discs on
the mounted needle

Experiment 13. Discussion

The colour of beetroot is due to the presence of a red pigment in the cell sap.

1 What evidence is there from your experiment that the cell wall and/or the cytoplasm is selectively permeable with respect to the pigment?

2 Which of the structures labelled in Fig. 3 contains (a) cellulose, (b) protein?

3 In what way is protein affected by temperatures above 50 °C?

4 What information about the selective permeability of the cell can be obtained from tubes 70 and 80?

5 How do your results support the hypothesis that it is the cytoplasm or part of the cytoplasm rather than the cell wall which is selectively permeable?

6 The thermal death point of cytoplasm is usually given as 45-50 °C. Why do you think the temperature in the experiment had to exceed 60 °C before the cell's selective permeability was markedly affected?

7 Apart from considerations of selective permeability, are there any other possible explanations for the final appearance of the liquid in tubes 70 and 80?

8 Why was it necessary to wash the discs thoroughly at the beginning of the experiment?

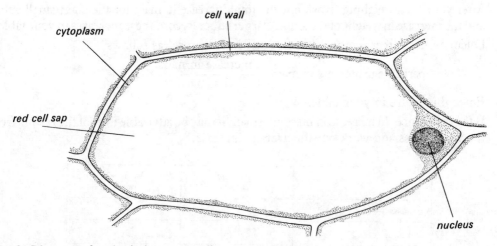

Fig. 3. Diagram of a single beetroot cell

Experiment 14. Osmosis and surface area*

(a) Use a scalpel or kitchen knife to cut eight cubes of potato with sides 1 cm long (i.e. each cube has a volume of 1 cm³). DO NOT cut on to the surface of the bench but use a tile or dissecting board.

(b) Similarly, cut one cube of potato with sides of 2 cm (i.e. the cube has a volume of $2 \times 2 \times 2 = 8$ cm³).

(c) Weigh the eight 1 cm³ pieces together and note their total mass.

(d) Weigh the 8 cm³ piece and note its mass.

(e) Place all the cubes in a beaker of water and note the time.

(f) In your notebook draw a table like the one below and enter the two masses in the appropriate column.

(g) After about one hour, remove the potato cubes from the water, dry them on a piece of blotting paper and weigh them. (As before, weigh the eight 1 cm³ pieces together.) Record the new masses in your table.

(h) If possible, return the cubes of potato to the water and repeat the weighings on the following day.

(i) After your final weighing, work out the total increase in mass for the eight small cubes and the increase in weight of the single, larger cube. Record these increases in your table.

(j) Calculate the percentage increase in mass in each case, as follows

$$\text{percentage increase in mass} = \frac{\text{increase in mass}}{\text{first mass}} \times 100$$

(k) Enter this figure in your table.

(l) If the differences in increase in mass are small, it may be advisable to pool the results from the whole class and work out the average increase.

	Eight 1 cm³ cubes	One 8 cm³ cube
1st mass		
2nd mass		
Increase in mass		
Percentage increase in mass		

*This experiment replaces the one on red blood cells, since blood sampling by teachers or pupils is no longer permitted. The experiment is based on one described by D. R. B. Barrett in 'Journal of Biological Education' 18 (4) 273.

Experiment 14.   Discussion

1 At the start of the experiment, by how much did the total mass of the eight 1 cm³ cubes differ from the mass of the 8 cm³ cube?

2 What is the most likely reason for this difference?

3 Would you expect there to be much difference in the total volume of the eight 1 cm³ cubes and the volume of the single 8 cm³ cube?

4 (a) What was the total surface area of the eight 1 cm³ cubes?
   (b) What was the surface area of the single 8 cm³ cube?

5 Why, do you suppose, the mass of the cubes increased after one hour's immersion in water?

6 What difference was there in the percentage increase in mass between the eight 1 cm³ cubes and the single 8 cm³ cube?

7 What is the most likely explanation of this difference?

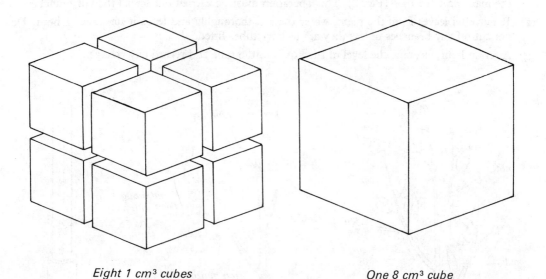

Eight 1 cm³ cubes   *One 8 cm³ cube*

Experiment 15.   Root pressure

You are provided with a potted plant. Read all instruction **(a)** before starting.

(a) Refer to Fig. 1 and then cut off the top of the main shoot and those branches which are sufficiently thick and sufficiently vertical to be used in the experiment. Make the cut just below a leaf at a point where the stem has a diameter of 5-7 mm. Leave the narrow or horizontal branches intact. Put all the cuttings in water.

(b) From the two different sizes of rubber tubing offered, select a piece which will fit firmly over one of the stumps without damaging it. Note which size tubing you have used.

(c) Use a dropping pipette to fill the rubber tubing attached to the branch, with water. Look to see if there is any leakage between the stem and rubber tubing. If there is, you will have to cut the stem lower down at a point where it is wider, or use a narrower piece of tubing.

(d) Select a piece of capillary tubing *from the same box* as the rubber tubing and insert it, marked end downwards, into the rubber tubing until it touches the stump. This will expel some liquid and air bubbles from the tube.

(e) Partially withdraw the glass tube from the rubber tubing until the liquid level corresponds to the mark near the base (Fig. 2). This operation must be carried out for all the cut branches.

(f) Remove all leaves from the plant, water the soil thoroughly and leave it for about 1 hour. Do not cut off any branches unless they are to have tubes fitted.

(g) After an hour, examine the level of the liquid in the tube and record any change.

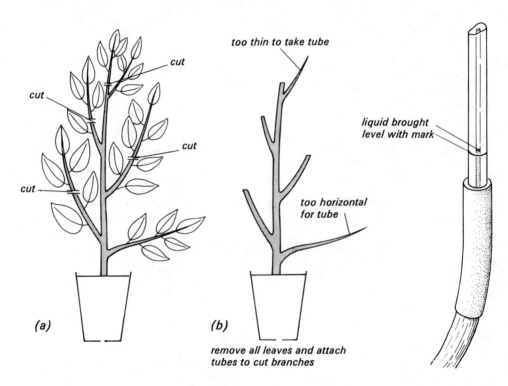

(a)

(b)

remove all leaves and attach
tubes to cut branches

too thin to take tube

cut

cut

cut

cut

too horizontal
for tube

liquid brought
level with mark

Fig. 1

Fig. 2

7 Transport in Plants

Introduction

Water, dissolved salts, food substances, 'hormones', oxygen and carbon dioxide are some of the substances which move about in plants. Some of them, e.g. the gases, move passively by simple diffusion, while others, e.g. food, are moved by active living processes. In school laboratories, it is difficult to investigate the movement of mineral salts and food in plants which is why there is only one experiment in this section on salt uptake (Experiment 15, p.222) and none at all on transport of food. In advanced laboratories, the movement of these substances is often followed by using radioactive compounds, a technique not available in most schools.

Most of the experiments described here are concerned with the passage of water through a plant and the exchange of gases with the atmosphere because these are fairly easy to demonstrate and measure.

Other experiments on transport in plants may be found in other sections of this book, for example Experiment 15, p. 196 'Root pressure'; Experiment 4a, p. 114 'Gaseous exchange in leaves'; Experiment 12, p. 156 'The effect of indole acetic acid on coleoptiles'.

Contents

Experiment 1.   Uptake and evaporation in leaves

(a) With a marker pen, label four test-tubes 1-4 and draw a line round each tube about 1 cm from the rim.

(b) Fill each tube up to the line with tap-water and place the tubes in a rack.

(c) Collect four leaves, as nearly as possible the same size.

(d) Place one of the leaves in tube 1 so that its leaf stalk is below the water level. (Fig. 1). Treat the other leaves with vaseline as follows:

(e) Place a leaf on a sheet of newspaper. Smear a thin layer of vaseline all over the *upper* surface of the leaf. Place this leaf in tube 2.

(f) Repeat this procedure with leaf number 3 but smear the vaseline on only the *lower* surface.

(g) Repeat this procedure with leaf number 4, but smear *both* surfaces with vaseline.

(h) Place the rack of tubes and their leaves in a position where they will not receive direct sunlight, and leave them for a week.

(i) Copy the table below into your notebook.

(j) From time to time during the week, check that the water level does not fall below the leaf stalk. If necessary, top up the tubes with water from a 2 cm^3 syringe and make a note in your table of the volume added.

(k) At the end of the week, use a 2 cm^3 syringe to add water to each tube and bring the water level up to the mark. In your table write the volume of water added in each case. If you added water during the week as well, work out the total volume of water added.

NOTE. The reading on the syringe measures the amount of water left in the barrel. The volume of water added to the tube is this figure taken away from 2 (Fig. 2).

Date set up	Volume of water added			
	1	*2*	*3*	*4*
Total				

Experiment 1.   Discussion

1 (a) In what way would you expect a layer of vaseline to affect evaporation from a leaf surface?
   (b) How might the vaseline produce this effect?

2 Which of the four leaves (a) took up least water, (b) evaporated least water? What assumptions are you making in your answer to (b)?

3 (a) What difference in water loss was there between leaves 2 and 3?
   (b) Suggest an explanation for this difference.

4 Would you expect the upper or lower surface of leaf number 1 to lose the greater amount of water? Explain your answer.

5 Water will evaporate directly from the exposed water surface in each tube. In what way will this affect the result?

6 Why was it necessary to use leaves of similar size?

7 Four leaves of similar size were placed in test-tubes A–D, exactly as described for this experiment, but no vaseline was used on any of them. After one week the tubes had lost water as follows: A. 6.4, B. 5.6, C. 8.0, D. 6.0 cm^3. How does this result affect the interpretation (a) of your own results, (b) of the combined class results?

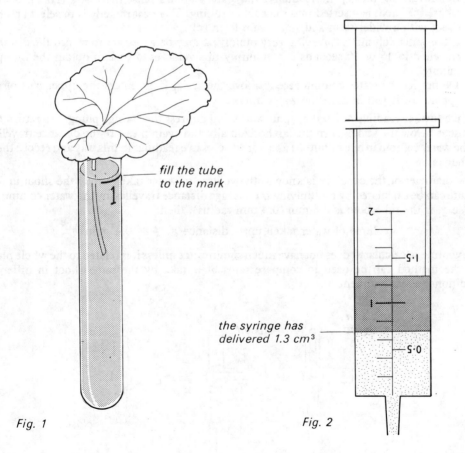

fill the tube
to the mark

the syringe has
delivered 1.3 cm^3

Fig. 1

Fig. 2

Experiment 2.　Uptake of water by shoots

(a) Copy the table on p. 203 into your notebook.

(b) Half fill a small jar with tap water; fill the syringe from this and expel all air bubbles.

(c) Fit the syringe to the side arm of the 3-way tap. Turn the tap down and press the syringe plunger gently until water wells out of the rubber tubing.

(d) Select a shoot and push the stem into the rubber tubing as far as possible.

(e) Turn the tap upwards and depress the syringe plunger to expel all air bubbles from the capillary.

(f) With the tap still upwards, withdraw the syringe plunger slightly to bring the water column to just below the start of the scale (Fig. 2).
Unless readings are being made in the field, the apparatus can be clamped in a retort stand at this stage.

(g) Turn the tap horizontally and watch the meniscus (air/water boundary) in the capillary. As the meniscus reaches the beginning of the scale note the time (or start the stop-clock) and after one minute notice how far the water has travelled up the capillary. Record this figure, in mm, in your table.

　(i) If, on turning the tap horizontally, the water column falls, there is a leak, probably where the shoot is inserted into the rubber tubing. The best remedy is to select another shoot with a wider stem and start again from (c).

　(ii) If the water column is travelling very rapidly, it may be necessary to record the distance travelled in 15 or 30 seconds and multiply this value by 4 or 2 to obtain the rate per minute.

　(iii) Do not let the water column recede above the scale. If air enters the upper part of the apparatus, it will impede the water uptake.

(h) To take further readings, turn the tap upwards and depress the syringe plunger to return the meniscus below the scale. Turn the tap horizontally and again measure the distance travelled by the water column in one minute. Take 3 or 4 successive readings in this way and record them in your table.

(i) If the diameter of the capillary is known, the volume of water taken up by the shoot in one minute can be calculated by multiplying the average distance travelled by the water column by πr^2, e.g. if the tube bore is 0.8 mm (0.4 mm radius), then:

$$\text{volume of water taken up} = \text{distance} \times \frac{22}{7} \times 0.4^2 \text{ mm}^3.$$

The volume so calculated does not have much significance unless it is related to the whole plant but the method can be used to compare rates of uptake by the same shoot in different conditions (*see* Experiment 3).

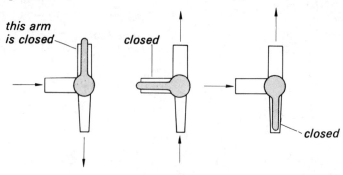

this arm is closed　　*closed*　　*closed*　　*closed*

Fig. 1
The 3-way tap

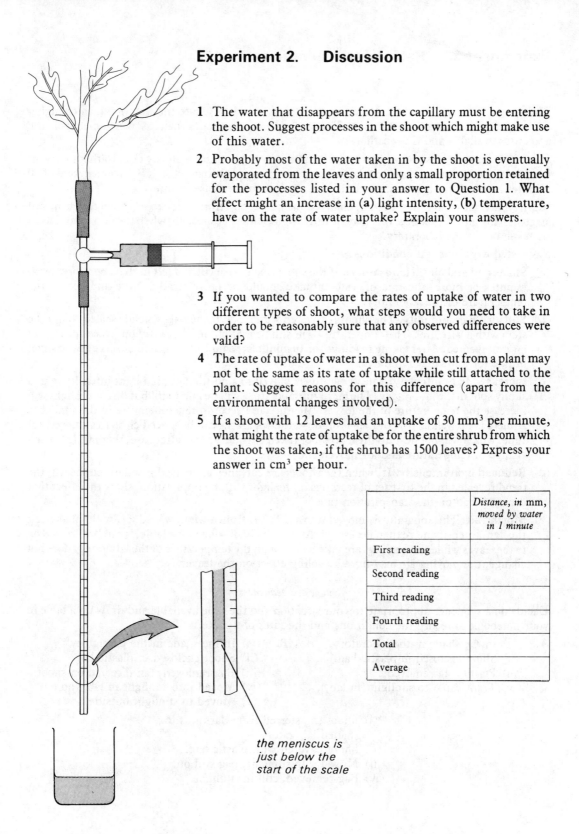

Experiment 2. Discussion

1 The water that disappears from the capillary must be entering the shoot. Suggest processes in the shoot which might make use of this water.

2 Probably most of the water taken in by the shoot is eventually evaporated from the leaves and only a small proportion retained for the processes listed in your answer to Question 1. What effect might an increase in (a) light intensity, (b) temperature, have on the rate of water uptake? Explain your answers.

3 If you wanted to compare the rates of uptake of water in two different types of shoot, what steps would you need to take in order to be reasonably sure that any observed differences were valid?

4 The rate of uptake of water in a shoot when cut from a plant may not be the same as its rate of uptake while still attached to the plant. Suggest reasons for this difference (apart from the environmental changes involved).

5 If a shoot with 12 leaves had an uptake of 30 mm³ per minute, what might the rate of uptake be for the entire shrub from which the shoot was taken, if the shrub has 1500 leaves? Express your answer in cm³ per hour.

	Distance, in mm, moved by water in 1 minute
First reading	
Second reading	
Third reading	
Fourth reading	
Total	
Average	

the meniscus is just below the start of the scale

Experiment 3.   Rates of transpiration

The apparatus and method for Experiment 2 are used to investigate the way in which changing conditions affect transpiration rates. The various experimental situations will depend on the apparatus available and the weather.

After each change, the shoot will take 15-30 minutes to acquire a new, steady rate. Rather than wait for this length of time it is best to take as many one-minute readings as possible over a period of 10 minutes and look for a consistent trend, i.e. increase or decrease in rate.

Ideally, only one condition should be changed each time, e.g. from shade to sun, cool to warm, still air to moving air. In practice, this may be difficult to achieve, especially when the apparatus has to be taken out of the laboratory.

Suggested experimental conditions.

(a) **Shade and still air (in laboratory).** If the shoots have been cut and left in the laboratory for 30 minutes or more, the rate of water uptake should be fairly steady. Note and record the temperature of the laboratory.

(b) **Still air, high humidity.** The shoot is covered with a polythene bag which is sealed round the stem with a wire 'tie'. The temperature will still be that of the laboratory but free exchange of air is prevented. After taking readings for 10 minutes, remove the bag and observe any changes in rate for a further 5 minutes.

(c) **Increased light intensity.** The apparatus is moved to a position of high light intensity, e.g. a sunny spot in the laboratory. The air in the laboratory will be fairly still but direct sunlight will increase the temperature of the leaves. Measure and record the temperature in the sunlight. If there is no well lit part of the laboratory, the apparatus can be placed close to a fluorescent tube (which will not heat up the leaves), or taken outside. In the latter case, there will also be a change in temperature and air movement.

(d) **Reduced light intensity.** If, when the shoots are cut, they are placed in a dark cupboard, the trend revealed in the first set of readings in the light will, by implication, show the effect that darkness had on the transpiration rate.

(e) **Moving air.** The apparatus is moved into a draught from a window or door, without altering the light intensity more than necessary. Alternatively, if a fan is available, it can be directed on to the leaves while the readings are taken. Although the temperature of the laboratory does not change, the moving air may have a cooling effect on the leaves.

Suggested sequences

Decide on a sequence appropriate to your situation and the time available and draw up a table in your notebook to record the conditions and the rates of uptake.

A.
(i) In the shade in the laboratory.
(ii) Shoot enclosed in plastic bag.
(iii) Plastic bag removed.
(iv) Transferred to sunlight (in lab.).

B.
(i) In the shade in the laboratory.
(ii) Shoot enclosed in plastic bag.
(iii) Bag removed, fan directed on shoot.
(iv) Moved into sunlight in lab. (no fan).
(v) Moved to sunlight outside.

C.
(i) Shoot transferred from darkness to shaded laboratory.
(ii) Shoot covered with plastic bag.
(iii) Moved into sunlight, bag still on.
(iv) Bag removed, still in sunlight.

Experiment 3.   Discussion

1 Which changes of conditions resulted in (a) an increase and (b) a decrease in the rate of water uptake?

2 Suggest reasons for the changes of rate in each case.

3 Of the conditions which caused an increased rate, which one seemed to have the greatest effect?

4 What changes in conditions are involved in taking a shoot from the shade in a laboratory to bright sunlight outside?

5 When the shoot was covered by a plastic bag, what effect, apart from a change in the rate of water uptake, was noticeable? What significance do you attach to this observation?

6 (a) How might an increase in light intensity affect the rate of photosynthesis in the shoot?
   (b) What influence might this change in photosynthetic rate have on the rate of water uptake?
   (c) On what grounds could you justify the assumption that a change in uptake resulting from an increase in light intensity is due largely to a change in the rate of transpiration rather than in the rate of photosynthesis?

7 It is usually emphasized that the potometer does not measure the rate of transpiration but only the rate of uptake.
   (a) Why do you think this distinction is made?
   (b) Why is it still reasonable to use the potometer to *compare* rates of transpiration of the same shoot in different conditions?

Experiment 4.   Rate of transpiration and water uptake

(a) Half fill a small jar with water; fill a 1 cm³ syringe from this and expel all air bubbles.

(b) Fit the syringe to the side-arm of the 3-way tap. Turn the tap down and depress the syringe plunger gently until water wells out of the rubber tubing.

(c) Select a shoot and push the stem into the rubber tubing as far as possible. Blot off any water remaining round the junction of the stem and tubing.

(d) Clamp the apparatus upright in a retort stand so that the end of the capillary is in the jar and below the water level.

(e) Turn the tap upwards and depress the syringe plunger to expel all air bubbles from the capillary.

(f) Withdraw the syringe plunger to refill the syringe and bring the plunger exactly to the 1.0 cm³ mark, i.e. the uppermost mark.

(g) Read all of this section and do not proceed unless the weighing can be made immediately after taking the apparatus from the water.
Turn the tap horizontally and take the apparatus out of the water; blot off any water drops clinging to the end of the capillary. Note the time and weigh the apparatus as follows:
   (i) If a top-pan balance is to be used, the apparatus may be placed horizontally across the pan, making sure that no part of the apparatus or shoot is touching the surroundings.
   (ii) If a beam balance is used, the apparatus is suspended by a wire loop from the hook holding one of the scale pans.
   Record the weight.

(h) During the next 10-15 minutes or more, clamp the apparatus upright but do not return the capillary to the water in the jar. Check that the water column in the capillary is receding and return it to near the bottom of the capillary from time to time. Turn the tap upwards, depress the syringe plunger carefully to send the water column down and then turn the tap horizontally again. Do not let air enter the upper part of the apparatus or it will impede the water uptake.
If, when the tap is horizontal, water escapes from the capillary, there is a leak, probably where the shoot fits into the rubber tubing. The simplest remedy is to change the shoot for one with a thicker stem and start again.

(i) Copy the table below into your notebook, but keep checking the water column and returning it to near the bottom of the capillary.

(j) After 10, 15 or 20 minutes (the longer the better), weigh the shoot and apparatus again. Record the new weight in your table.
Now turn the tap upwards and depress the syringe plunger to return the water column exactly to the bottom of the capillary. Note and record the volume reading on the syringe.

(i) *First weight* in grams	(iii) *First volume* reading in cm³	1.0
(ii) *Second weight* in grams	(iv) *Second volume* reading in cm³	
Weight of water transpired in ... minutes (i) − (ii)	Volume of water taken up in ... minutes (iii) − (iv)	

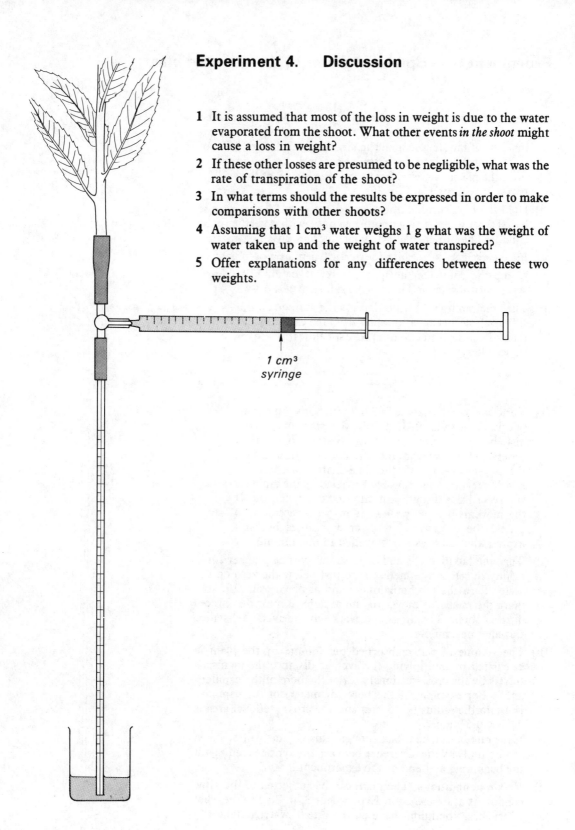

Experiment 4.   Discussion

1 It is assumed that most of the loss in weight is due to the water evaporated from the shoot. What other events *in the shoot* might cause a loss in weight?

2 If these other losses are presumed to be negligible, what was the rate of transpiration of the shoot?

3 In what terms should the results be expressed in order to make comparisons with other shoots?

4 Assuming that 1 cm³ water weighs 1 g what was the weight of water taken up and the weight of water transpired?

5 Offer explanations for any differences between these two weights.

1 cm³
syringe

Experiment 5.   Uptake of water by an uprooted plant

(a) Select an uprooted plant and carefully dry the lower part of the stem. Mould a ball of 'Blu-tack', or 'Plasticine' closely round this part of the stem to make an air-tight seal. Lower the root system carefully into the barrel of a 5 or 10 cm³ syringe and press the plastic material firmly into the mouth of the syringe barrel to seal the plant in (Fig. 1).

(b) Hold the plant upside down and completely fill the syringe barrel with water, using another syringe with a needle fitted (Fig.2). The presence of a few air bubbles does not matter.

(c) Fit the syringe barrel into the vertical arm of a potometer.

(d) Clamp the potometer upright in a retort stand. Fill a small syringe with water and fit it to the side-arm of the 3-way tap.

(e) Turn the tap upwards and depress the syringe on the side arm to expel all air bubbles from the capillary and then adjust the water column to just below the zero on the capillary scale.

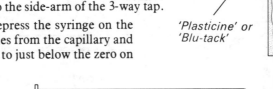

'Plasticine' or 'Blu-tack'

(f) Turn the tap horizontally and watch the meniscus (air/water boundary). If it falls, the plant is not adequately sealed into the syringe barrel. If it rises, observe when it reaches the zero on the scale and start the stop-clock (or note the time). After one minute, note how far the meniscus has travelled up the capillary and record this distance, in mm, in your notebook. If the movement is very slow, it may be necessary to extend the observation to 2 or 3 minutes but still express the result as mm travelled in one minute.

(g) Turn the tap upwards and depress the syringe plunger sufficiently to return the meniscus to just below the zero on the scale. Turn the tap horizontally and again measure the distance the meniscus moves in one minute. Repeat this procedure to obtain 3 readings and work out the average distance travelled per minute.

(h) The volume of water absorbed per minute by the plant is calculated by multiplying the average distance the meniscus travels by the cross-sectional area of the bore of the capillary (πr^2). For example, if the bore (diameter) of the capillary is 1 mm the radius is 0.5 mm and the cross-sectional area is $22/7 \times 0.5^2$ mm².
   If the rate of uptake is fast enough, this volume will be given more simply by the difference between the syringe readings at the beginning and end of the experiment.

(i) **Varying conditions.** The plant can be subjected to the same conditions as described in Experiment 3 (p. 204) to see what effect these conditions have on the rate of water uptake.

Fig. 1

Experiment 5.   Discussion

If you have already answered the questions 1–3 on p. 203 (Experiment 2) the first 3 questions of this discussion can be ignored.

1 The water that disappears from the capillary must be entering the plant. Suggest processes in the shoot which might make use of this water.

2 Probably most of the water taken in by the plant is eventually evaporated from the leaves and only a small proportion retained for the processes listed in your answer to Question 1. What effect might an increase in (a) light intensity, (b) temperature have on the rate of water uptake? Explain your answers.

3 If you wanted to compare the rate of uptake of water in two different types of plant, what steps would you need to take in order to be reasonably sure that any observed differences were valid?

4 The rate of water uptake from an uprooted plant may not be the same as when the plant was in the soil. Suggest reasons for any differences.

5 The large volume of water in the 10 cm³ syringe barrel, attached to a capillary, must function like a thermometer. What effect will this have on the results (a) if the apparatus was kept in the same situation, (b) if the apparatus was moved to a variety of situations?
Suggest a simple control to help eliminate this source of error.

6 If you made measurements of water uptake in different situations, answer the questions on p. 205 (Experiment 3) unless you have already done so.

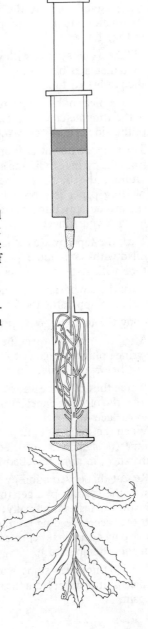

	Distance travelled by meniscus in 1 minute
First reading	mm
Second reading	mm
Third reading	mm
Average	mm

Fig. 2

Experiment 6.   Conditions affecting evaporation

The apparatus, called an atmometer, represents some of the physical attributes of a leaf in so far as simple evaporation is concerned.

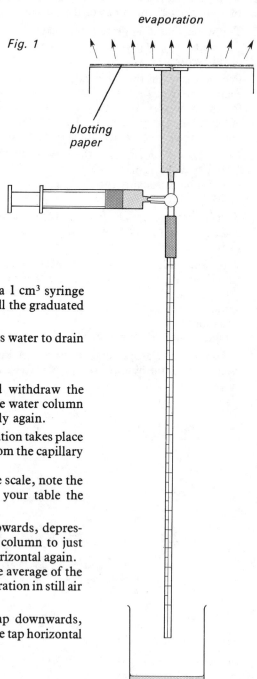

Fig. 1

evaporation

blotting paper

(a) Clamp the graduated tube upright in a clamp stand and fit the nozzle of the atmometer into the top of the 3-way tap as in Fig. 1.

(b) Fill the 5 cm³ syringe with water from the container and place the container below the graduated tube.

(c) Place a piece of blotting paper on the disc of the atmometer and fit the 5 cm³ syringe to the side arm of the 3-way tap.

(d) Turn the tap downwards and inject water from the syringe to fill the atmometer and saturate the blotting paper. When the blotting paper is thoroughly moistened, smooth out any bubbles trapped beneath it, with your finger.

(e) Turn the tap upwards, change the 5 cm³ for a 1 cm³ syringe filled with water and depress the plunger to fill the graduated tube with water and expel any air bubbles.

(f) Turn the tap horizontally and allow the surplus water to drain from the apparatus for 5 minutes.

(g) Copy the table opposite into your notebook.

(h) After 5 minutes, turn the tap upwards and withdraw the syringe plunger slightly to bring the end of the water column just below the scale. Turn the tap horizontally again.

(i) Watch the end of the water column. As evaporation takes place from the blotting paper, water is withdrawn from the capillary to replace it.
When the water column passes the zero on the scale, note the time (or start the stop-clock) and record in your table the distance, in mm, travelled in 1 minute.

(j) Repeat this observation by turning the tap upwards, depressing the syringe plunger to return the water column to just below the scale and, finally, turning the tap horizontal again.
If the second reading is similar to the first, the average of the two may be taken to represent the rate of evaporation in still air at room temperature.

If the atmometer stops working, turn the tap downwards, inject a little water from the syringe and turn the tap horizontal again.

(k) The conditions can now be varied and the rates of evaporation compared. Take enough readings in each case to obtain consistent results.

    (i) *Air movement*. The apparatus is placed in front of a fan in the laboratory or simply fanned with an exercise book.

    (ii) *Temperature*. A 60 watt lamp bulb, in a bench lamp for example, supported about 15 cm above the blotting paper will raise its temperature by about 15 °C.

    (iii) *Light intensity*. It is difficult to increase light intensity without at the same time raising the temperature of the atmometer as in (ii) above or when the apparatus is moved into a patch of sunlight in the laboratory.
However, if a fluorescent strip-light is available, the apparatus can be placed 15 cm below this without significantly raising its temperature.

    (iv) *Humidity*. A Petri dish base with perforations is placed over the blotting paper with its base uppermost.

Distance in mm travelled by water column in 1 minute				
Still air	*Moving air*	*Increased temperature*	*Increased light intensity*	*Increased humidity*

Experiment 6.   Discussion

1 State the conditions which increased or decreased the rate of evaporation. Discuss whether it is possible to decide on their relative importance.

2 If you have done Experiment 3, discuss the extent to which the conditions which influenced the rate of evaporation also affected the rate of transpiration in the living shoot. Try to account for any differences in the results.

3 In what ways does the living shoot differ from the atmometer with regard to structural features which might affect the rate of evaporation?

Experiment 7. Water tension in the stem

(a) Clamp the potometer upright in a retort stand. Fill a small syringe with water, expel all air bubbles and fit it to the side arm of the 3-way tap.

(b) Turn the tap down and depress the syringe plunger until water emerges from the rubber tubing.

(c) Insert the stem of a cut shoot into the rubber tubing. It is essential to make a good seal.

(d) Lower the potometer so that the capillary tube dips into a small container of water with some mercury at the bottom. The end of the capillary must be below the mercury (Fig. 1).

(e) Turn the tap upwards and depress the syringe to expel all air bubbles from the capillary.

(f) Turn the tap horizontally and observe the apparatus from time to time over a period of 30-45 minutes. At the end of this time measure the height of mercury above the mercury level in the reservoir and record this distance in your notebook.

Experiment 7. Discussion

1 Atmospheric pressure is helping to push the mercury up the capillary but before it can do this the shoot must reduce the water pressure in the capillary to below atmospheric pressure. From your results calculate the reduction of pressure, in atmospheres, in the stem. (A column of mercury 760 mm high exerts a pressure of 1 atmosphere.)

2 Atmospheric pressure can support a column of water about 10 metres high. Many trees are taller than this. How could transpiration account for the movement of water in these trees?

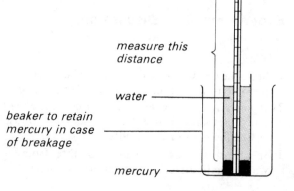

Fig. 1

measure this distance

water

beaker to retain mercury in case of breakage

mercury

Experiment 8.   Pathways for gases in a leaf

(a) If you have not already tried Experiment 9, or do not plan to do it, paint an area about 10 mm square on both the upper and lower surface of a leaf with a thin layer of clear nail varnish and leave the varnish to dry for 10 minutes. Meanwhile, continue with instruction (b).

(b) Fill a beaker two thirds full with water (hot water if available) and heat it on a tripod and gauze over a Bunsen burner until it reaches about 70 °C.

(c) Extinguish the burner when the water is hot enough.

(d) Hold a leaf (not the one with the nail varnish) in forceps and plunge it into the hot water (Fig. 1), whilst observing the lower surface of the leaf.

(e) Repeat the experiment with a fresh leaf but this time watch the upper surface.

(f) If you followed instruction (a), use fine forceps to peel the dried varnish from the lower surface, place it on a slide and examine it under the microscope. Only a small piece of peel is needed.

(g) Count the number of stomata visible in the field of the microscope and record the results in your notebook. If the stomata are too numerous to count over the whole field of vision, count only those in, say, a quarter of the field or between marks on the slide. Alternatively, use a higher magnification if available.

(h) Repeat the operation with the nail varnish from the upper surface.

Experiment 8.   Discussion

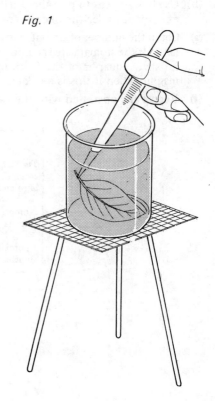

Fig. 1

1 What did you observe when the leaf was placed in hot water while watching (a) the lower surface, (b) the upper surface?

2 What was the function of the hot water in this experiment?

3 Judging from your varnish peels, which surface of the leaf had the greater number of stomata?

4 Use your results and your knowledge of leaf anatomy to explain your observations in this experiment.

Experiment 9.   Evaporation from the leaf surface

Cobalt chloride paper is pink when moist and blue when dry. It must be handled with forceps because the moisture from your fingers will turn it pink.

If the paper is not deep blue when you start the experiment, place it on the warm shade of a bench lamp (switched on) for a few seconds.

(a) Cut two pieces of adhesive transparent tape about 20 mm square.

(b) Cut a fairly large leaf from the plant or shoot provided.

(c) Use forceps to collect two small pieces of cobalt chloride paper and place each one in the centre of the sticky surface of the adhesive tape squares.

IN THE NEXT OPERATION, AVOID PRESSING THE LEAF SURFACE AGAINST THE COBALT CHLORIDE PAPER

(d) Note the time and stick one of the squares of tape with the cobalt chloride paper to the upper surface of the leaf blade, avoiding the midrib and other prominent veins. Stick the other square on the lower surface, opposite to the first one. Press the edges of the tape to make a good seal but do not press the leaf against the cobalt chloride paper (Fig. 1).

(e) While waiting for the colour change, paint an area about 10 mm square on the upper surface of the leaf with a thin layer of clear nail varnish. Do the same on the lower surface and leave the varnish to dry for 10 minutes.

(f) Copy the table below into your notebook.

(g) Record the time taken for each piece of cobalt chloride paper to become pink. This may be difficult to judge but it is only a *comparison* of results that is needed.

(h) Use fine forceps to peel the dried varnish from the lower leaf surface, place it on a slide and examine it under the microscope (Fig. 2). Only a small piece of peel is needed.

(i) Count the number of stomata visible in the field of the microscope and record the result in your table. If the stomata are too numerous to count over the whole field of vision, count only those in, say, a quarter of the field, or between marks on the slide. Alternatively, use a higher magnification if this is available.

(j) Repeat the operation with the varnish peel from the upper surface.

	Upper surface	Lower surface
Time taken for cobalt chloride to go pink.		
Number of stomata		
Approximate % stomata open		

Experiment 9.   Discussion

1   On which surface of the leaf did the cobalt chloride paper go pink first?

2   What does this indicate about the relative rates of evaporation from each surface?

3   From your observations of the nail varnish peels, which surface of the leaf had the greater number of stomata?

4   How can you relate the difference in the number of stomata on each surface to the difference in the rates of evaporation?

5   From your peel, could you tell whether the stomata were open or closed?

6   If the stomata were closed, how would you expect this to alter the result as described in your answer to Question 1?

7   If the stomata had all been closed, but the results with cobalt chloride were the same as in Question 1, use your knowledge of the anatomy of a leaf to offer an alternative explanation for Question 4.

8   What part does the sticky tape play, apart from holding the cobalt chloride paper on the leaf?

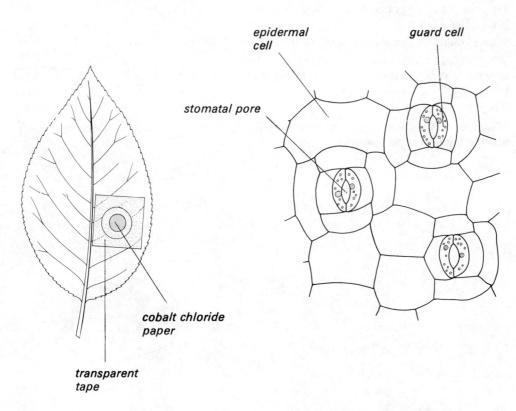

Fig. 1

Fig. 2   Appearance of stomata in leaf epidermis (× 200)

215

Experiment 10.   To collect and identify the product of transpiration

(a) You are provided with a potted plant. Water the soil if it appears to be dry.

(b) Place a transparent plastic bag over the entire shoot and close the mouth of the bag firmly round the base of the stem with a wire tie or something similar (Fig. 1).

(c) Place the plant under an artificial light-source or in a position where it can receive sunlight.

(d) After several hours in the light, note any changes of conditions inside the plastic bag.

(e) Remove the plastic bag carefully, hold it with the open end upward and shake the collected liquid into one corner.

(f) Remove a sample of the liquid with a clean dropping pipette and test it by dropping it on to some anhydrous copper sulphate or blue cobalt chloride paper.

Experiment 10.   Discussion

1 Describe the changes that had taken place in the plastic bag.

2 Where could the liquid have come from?

3 What controls would have to be carried out to be reasonably sure about the source of this liquid?

4 What evidence is there that the liquid (a) contains water, (b) is pure water?

5 What tests would have to be carried out to verify that the liquid was pure water?

Fig. 1

Experiment 11.   To trace the path of water through the shoot

CARE. Methylene blue will stain your clothes. Wear a laboratory coat if possible and take care not to drop or flick the dye.

(a) Place a freshly cut shoot in a small quantity of methylene blue solution and leave it for 30 minutes in bright light.

(b) At intervals of about 10 minutes, without removing the shoot from the dye, study the stem and leaves to see if there is any evidence of movement of the dye. Note which leaves, if any, have the dye in their veins.

(c) After 30 minutes, or longer if necessary, examine the shoot again and note any changes in the distribution of the dye.

(d) Fill a beaker or jar with cold water for rinsing the dye off the stem.

(e) Remove the shoot from the methylene blue and carefully wash the dye from the stem, in the jar of water. With a razor blade or scalpel cut across the stem a few millimetres from the bottom (Fig. 1), avoiding the region which has become deeply stained with the dye. Examine the cut surface of the stem with a hand lens and make a simple outline diagram to show the distribution of dye in the stem.

(f) Select a leaf whose veins have been coloured by the dye. Use a razor blade to cut across the stalk and the leaf blade and examine the cut surfaces with your lens. Make drawings in your notebook to show the distribution of the dye in the leaf stalk and leaf blade.

(g) If the shoot is not needed for further experiments, the stem can be cut across at higher and higher levels to see how far the dye has travelled.

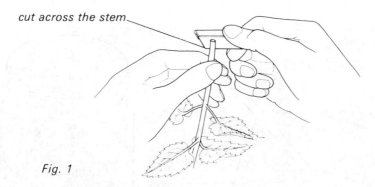

cut across the stem

Fig. 1

Experiment 11.   Discussion

1 What properties must a dye, such as methylene blue, have to be useful in this experiment?

2 It is assumed that the movement of dye through the shoot shows the normal route taken by water through a plant. What objections are there to this assumption?

3 Assuming that these objections can be dismissed, describe the path taken by water through the stem to the leaf as revealed by your experiment.

Experiment 12.   Conducting pathways in the shoot

(a) Fill the large container with water.

(b) Withdraw the syringe plunger to the top mark or beyond and fit the syringe to the cut end of the stem by means of the rubber tubing.

(c) Immerse the leafy shoot in the container of water and press the syringe plunger (Fig. 2). Keep the pressure as high as possible without breaking the seal between the rubber tubing and the syringe.

(d) Maintain the pressure for about 30 seconds and look for any signs of bubbles appearing from the stem or leaves.

(e) Cut all or most of the leaves transversely across in the mid-line (Fig. 1) to leave only the lower half of the leaf attached to the stem. Repeat the experiment, maintaining the pressure for at least 30 seconds.

(f) Now cut the leaves off completely, leaving only the petioles (leaf stalks) attached to the stem and repeat the experiment.

(g) Finally, cut a piece of stem about 5 cm long. Fill the syringe with water, connect it to the twig, point the apparatus vertically upwards and very slowly depress the plunger to see if it is possible to force water through the stem (Fig. 3).

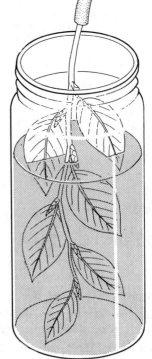

Fig. 2

air

Fig. 1

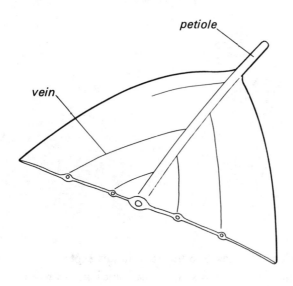

petiole

vein

Experiment 12.   Discussion

Although in this experiment, air is being forced through the shoot, it is assumed that it is following the pathway normally taken by water.

1 Did any air escape from the stomata of the leaves or the lenticels (pores in the bark) of the stem?
2 Did air escape (a) from the veins of the cut leaves (b) from the leaf stalks when the leaves were cut off?
3 Was it possible to force water through the stem?
4 What conclusions are you entitled to draw from the results of this experiment, about the conducting pathways in the shoot?

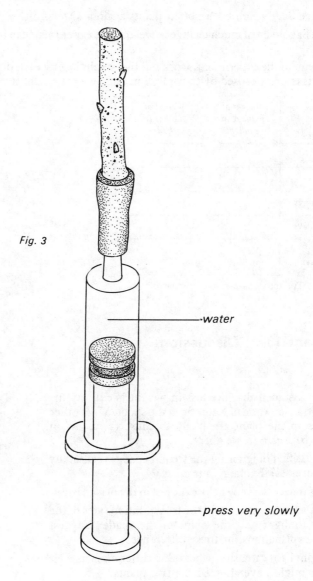

Fig. 3

water

press very slowly

Experiment 13. Measuring the transpiration rate of a potted plant

(a) Label one pot A, the other B and add your initials to each.

(b) Add sufficient water to each pot to make the soil moist.

(c) Copy the table below into your notebook.

(d) Place each pot in a plastic bag and tie the bag securely round the base of the plant's stem, using string or wire 'ties' (Fig. 1).

(e) Weigh each potted plant on a balance and record the weights in your table.

(f) Leave plant A in a sunny position in the laboratory and place plant B in a shaded position in the same room.

(g) After one day, weigh both potted plants again and record their weights in your table.

(h) If there has not been much change in weight, the experiment can be continued for several more days.

(i) At the end of the experiment, work out the weight lost by each plant and, assuming the loss in weight is entirely caused by transpiration calculate the rate of transpiration in grams per day.

Time	Weight of plant A in sunlight	Weight of plant B in shade
Start		
After 1 day		
After . . . . days		
After . . . . days		
Loss of weight in . . . .days		

Transpiration rate =   grams per day

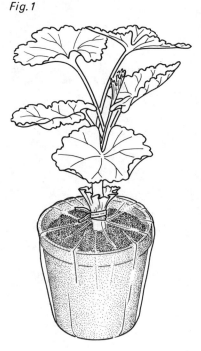

Fig. 1

Experiment 13. Discussion

1 We have assumed that the loss in weight is entirely the result of transpiration of water from the shoot. What other processes in the plant are likely to cause (a) a loss in weight, (b) a gain in weight?

2 Are we justified in ignoring the weight changes caused by these processes? Explain your answer.

3 Why was it necessary to enclose the pot in the plastic bag?

4 What difference was there in the transpiration rates of the plant in sunlight and the plant in the shade? Suggest possible explanations for these differences.

5 Apart from light intensity, what other differences in their positions might have affected the two plants?

Experiment 14.   Measuring the transpiration rate of an uprooted plant

(a) Label each bottle or flask with your initials.

(b) Fill each vessel with water to within 1 cm of the rim.

(c) Place the uprooted plant in one of the vessels.

(d) Copy the table below into your notebook.

(e) Weigh the vessel with the plant and record its weight in your table. Weigh the other vessel (with no plant) and record its weight.

(f) Leave both vessels in a (potentially) sunny position in the laboratory.

(g) After one day, weigh both vessels again and record their weights in your table.

(h) If there has been little change of weight in the vessel with the plant, the experiment can be continued for several more days.

(i) At the end of the experiment, work out the weight lost by each vessel. Subtract the weight lost by the vessel with water only, from the weight lost by the vessel with the plant. This will give the loss in weight resulting from the plant's transpiration.

(j) Calculate the rate of transpiration in grams per day.

Time	Weight of vessel with plant	Weight of vessel without plant
Start		
After 1 day		
After . . . days		
After . . . days		
Loss of weight		
Weight lost as a result of transpiration		Transpiration rate = g per day

Experiment 14.   Discussion

1 Why was it necessary to include a vessel with water but no plant?

2 Why was it necessary to deduct the weight lost by the vessel with no plant, from the weight lost by the vessel with a plant?

3 Would you expect the direct evaporation from each vessel to be the same? Explain your answer.

Experiment 15.   Uptake of nitrate and potassium by plants

(a) Copy the table below into your notebook.

(b) Three-quarters fill two test-tubes with a solution containing 1000 mg/l potassium and 500 mg/l nitrate(V) ions.

(c) Into each tube place an uprooted plant so that the root system is immersed in the solution. Select a pair of plants as similar as possible in size, leaf number and root system.

(d) Use a marker pen to mark the level of solution in each tube, label the tubes A and B and add your initials.

(e) Cover the entire shoot of plant B with a polythene bag and secure the mouth of the bag round the top of the test-tube with an elastic band.

(f) Place both plants in a position where they will receive daylight but not direct sunlight and inspect the solution level each day for a week.

(g) Top up the solution as required with *distilled water* to keep the level constant.

(h) At the end of the week, remove the polythene bag, top up both tubes to the mark with distilled water.

(i) Remove the plants, close the mouth of each tube in turn with your thumb and invert several times to obtain a homogeneous solution. Now test each solution for nitrate and potassium using Merckoquant test strips as follows:

Nitrate. Immerse the paper at the end of the plastic strip in solution A for about *2 seconds*. Leave the strip on the bench for *2 minutes* and then compare the colour of the lower test square with the colours on the side of the container. (The upper square should not change colour. If it goes mauve, it means that NO_2 ions are present and the nitrate(V) test is unreliable.)

Record in your table the level of nitrate present, in mg/l. Use a fresh strip to test the solution in tube B and record the result.

Potassium. Immerse the paper square at the end of the test strip in solution A for about *2 seconds*, shake off any excess and then place the end of the test strip in 0.1 M nitric acid for *1 minute* without swirling the strip about. At the end of this time, blot off the excess acid and compare the colour with those shown on the container. Record in your table the concentration of potassium left in the solution. Repeat the test for tube B with a fresh strip.

(j) *Control.* A pair of test tubes containing the solution but no plants will have been left for a week and kept topped up with distilled water. The solutions are tested as described above (one test for the whole class will be sufficient) and the concentrations of nitrate and potassium recorded in the table.

Experiment 15.   Discussion

1 What changes in the concentrations of nitrate and potassium occurred in tubes A and B after one week? Point out any differences between the two tubes and the two ions.

2 What effect would you expect the plastic bag to have on the plant's rate of transpiration, water uptake and photosynthesis? What other effects might the bag have on the plant's metabolism?

3 If the plant had taken up water but no salts during the week, what difference in the concentrations of potassium and nitrate would you expect after a week of topping up with distilled water?

4 If the rate of salt uptake was directly related to the rate of water uptake, what differences would you expect to find between solutions A and B?

5 If the rate of salt uptake was independent of the rate of water uptake, what differences would you expect to find between solutions A and B after a week?

6 Advance a simple hypothesis to account for the uptake of nitrate and potassium ions by the plant used in your experiment. What further experiments would you do to test this hypothesis?

	TUBE A Shoot not covered		TUBE B Shoot covered	
	Nitrate	Potassium	Nitrate	Potassium
CONTROL (no plant)				
PLANTS (after one week)				

8 Respiration and Gaseous Exchange

Introduction

To a biologist, the term 'respiration' means the series of chemical reactions taking place in the cells of living organisms, which make energy available for driving other reactions essential to life processes. The energy is derived from food and the reactions which release it for other purposes involve breaking down the molecules of food to carbon dioxide and water. In *aerobic respiration*, oxygen plays a part in the breakdown; in *anaerobic respiration* the breakdown takes place without oxygen.

An equation summarizing the process of aerobic respiration is given below.

$$C_6H_{12}O_6 \quad + \quad 6O_2 \quad = \quad 6CO_2 \quad + \quad 6H_2O \quad + \quad \text{Energy}$$

| glucose decrease in dry weight (utilization of food) | uptake of oxygen | output of carbon dioxide | e.g. heat production, movement. |

Thus, decrease in dry weight, uptake of oxygen, production of carbon dioxide or liberation of energy are considered to be evidence that respiration is taking place and most of the experiments in this book are concerned with seeking evidence of this kind.

The uptake of oxygen and output of carbon dioxide is called *gaseous exchange*. It should not be confused with respiration as defined above although, of course, it is used as evidence of respiration. Similarly, the filling and emptying of our lungs is called *ventilation*. It is necessary to ventilate the lungs to promote gaseous exchange. Experiment 4 is concerned with one aspect of ventilation. (Experiment 10, p. 64, Experiment 4, p. 114 and Experiment 14, p. 161 are also concerned with aspects of respiration.)

Contents

Experiment 1. The products of combustion of food

Cobalt chloride paper is blue when dry but goes pink in the presence of water or water vapour.

(a) Fill a beaker or jar two thirds full with tap water.

(b) Collect a jam jar with a lid, check that it is clean and dry and remove the lid.

READ INSTRUCTIONS **(c)** TO **(e)** BEFORE PROCEEDING.

(c) Place a small sample of food in the wire spoon provided, and hold it in a Bunsen flame until the food ignites and continues to burn steadily *when removed from the Bunsen flame*.

(d) Hold the dry jar upside down about 2 cm above the burning food so that the gases from the flames enter the jar.

(e) When the sides of the jar appear misty, place the wire spoon and burning food in the water, put the lid on the jar you are holding and place it upright on the bench.

(f) Use forceps to pick up a piece of cobalt chloride paper by one corner, hold it about 20 cm above a small Bunsen flame until it turns deep blue and then, briefly removing the lid from the jar, drop the cobalt chloride paper in and replace the lid.

(g) Turn the jar sideways so that the cobalt chloride paper comes in contact with the sides and leave it for one minute. Note any colour change in the cobalt chloride at the end of this time.

(h) Pour a little lime water into the jar, briefly removing the lid to do so. Replace the lid and shake the lime water in the jar. Note any change in its appearance.

(i) CONTROL. Wash the jar and lid and dry both thoroughly (or use a fresh jar and lid).

(j) Use forceps to pick up a piece of cobalt chloride paper and turn it blue as before. Drop the paper into the jar and replace the lid for one minute. Note any colour change.

(k) Shake a little lime water in the jar as before and note any change in its appearance.

(l) FURTHER EXPERIMENTS. The experiment may be repeated for other food samples to see if they all give similar results. The wire spoon can be renewed by removing the aluminium foil and charred remains of food and replacing it with a fresh piece of foil. The jar must be cleaned and thoroughly dried as before. There is no need to repeat the control.

Remove the lid briefly to put the cobalt chloride paper in the jar

Experiment 1.   Discussion

1 What gases are produced by the samples of burning food?

2 What evidence have you that these gases are produced?

3 Both these gases are already present in the atmosphere. How can you be sure that your tests relate to the gases from burning food and not those already present in the air?

4 Are these gases the only product of combustion of food?

5 What evidence is there that energy is released from the food during the experiment?

6 Assuming that all food substances contain carbon and hydrogen, explain the production of the two gases from burning food in simple chemical terms.

7 Living organisms derive their energy from food. In what ways is the burning of food **(a)** similar to, **(b)** different from, the use of food for providing energy in the body?

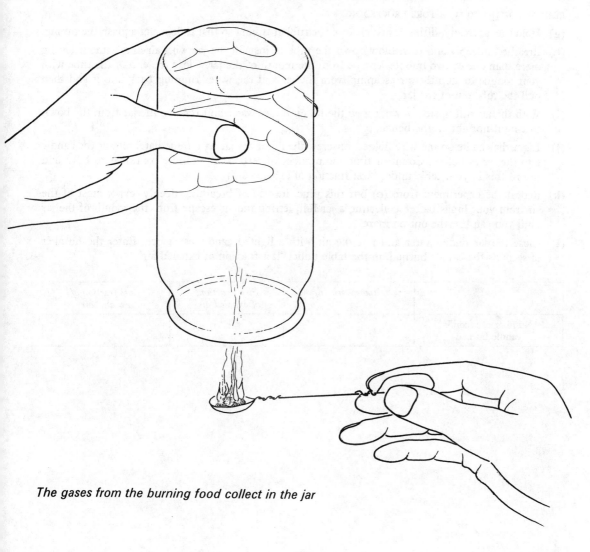

The gases from the burning food collect in the jar

Experiment 2.   Exhaled air (1)

(a) Copy the table given below into your notebook.

(b) Collect a screw-top jar with lid, and a candle on a special wire holder.

(c) Remove the lid from the jar and light the candle. When it is burning steadily, lower the candle into the jar till the lid is pressed firmly against the jar's rim (see Fig. 3). Count or time the number of seconds the candle continues to burn and record this figure in your table under 'atmospheric air'.

(d) Fill a bowl (or pneumatic trough) with just sufficient water to cover the jar when it is lying on its side.

(e) Place the jar on its side under the water and insert the rubber tube into the jar as shown in Fig. 1.

(f) Turn the jar upside down with the tubing still inside as in Fig. 2.

READ ALL OF **(g)** TO **(i)** BEFORE PROCEEDING.

(g) Hold the jar steady, lifting it just clear of the rubber tubing so that it does not squash the tubing.

(h) Breathe in deeply and then blow down the tube to just fill the jar with air and without letting more than one or two bubbles escape from the mouth of the jar. Close the end of the tube with your tongue to stop the air escaping from the jar and the water running back into it and then pull the tube out of the jar.

(i) With the jar still upside down, screw the lid on under water and remove the jar from the bowl, placing it upright on the bench.

(j) Light the candle on the wire holder, unscrew the lid of the jar and immediately lower the candle into the jar as before. Count or time the number of seconds the candle continues to burn and record this in your table under 'first fraction of exhaled air'.

(k) Repeat the experiment from **(e)** but this time, instead of breathing deeply, expel most of the air from your lungs before collecting a jar full, letting the air escape from the mouth of the jar until you can breathe out no more.

(l) Close the jar under water and test the air with a lighted candle as before. Enter the number of seconds the candle burned, in the table under 'last fraction of exhaled air'.

	Atmospheric air	First fraction of exhaled air	Last fraction of exhaled air
Number of seconds candle burned			

Experiment 2.   Discussion

1 Which atmospheric gas must be present if burning is to take place?

2 What do the results suggest about the differences in the amounts of this gas in (a) exhaled air and (b) ordinary atmospheric air?

3 What difference in results was there between the two samples of air, one collected from the early part of exhalation and one from the final stages?

4 Suggest possible explanations for this difference, considering which parts of the breathing system the air has come from.

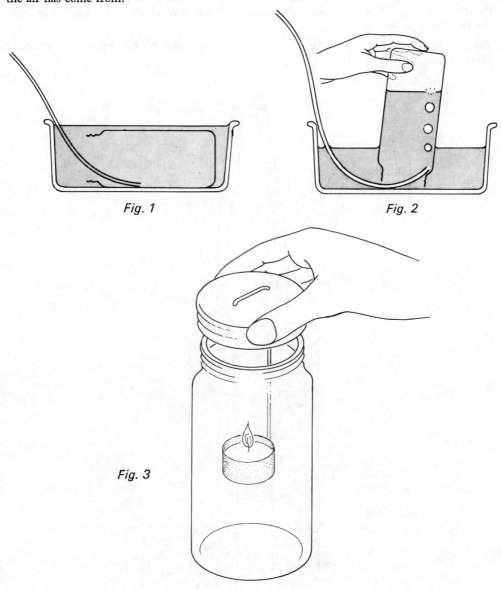

Fig. 1

Fig. 2

Fig. 3

Lower the burning candle into the jar until the lid is pressed against the rim

Experiment 3. Exhaled air (2)

You are provided with two large test-tubes fitted with rubber bungs and delivery tubes as shown in Fig. 1, and also two lengths of rubber tubing. Label the tubes A and B.

(a) Use a graduated pipette to place 10 cm^3 lime water in each test-tube.

(b) Place the bungs in both tubes and connect one length of rubber tubing to each test-tube as follows: with tube A fit the rubber tubing to the shorter glass tube, with tube B fit the rubber tubing to the longer glass tube (see Fig. 1).

(c) Put the ends of both rubber tubes at the same time in your mouth and breathe in and out through the tubes for about 15 seconds. Notice which tube is bubbling when you breathe out and which one bubbles when you breathe in.

(d) Compare the lime water in each tube. If there is no difference, breathe in and out through the tubes for another 15 seconds.

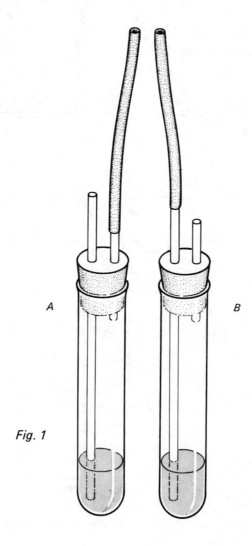

Fig. 1

Experiment 3.   Discussion

1. Where did the air passing through test-tube A come from? Where did the air passing through test-tube B come from?
2. What difference was there in the appearance of the lime water in the two tubes?
3. Which gas turns lime water milky?
4. What do the results tell you about the composition of the air you breathe out and the air you breathe in?
5. Why was it necessary to include tube A when results could have been obtained simply by breathing out through the lime water in B?

Breathe in and out through the tubing

Experiment 4. Vital capacity and tidal volume

Vital capacity

(a) Put water in the bowl to a depth of about 50 mm. Fill the container with water from the tap and put on the lid or stopper.

(b) Hold the container upside down in the bowl, with the lid or stopper under the water. Remove the stopper under water without letting air in. Some water will escape if the container is made of plastic but this does not matter.

(c) Push a length of rubber tubing through the mouth of the container to position A (Fig. 1). Take a deep breath and blow through the tube until you can exhale no more.

(d) All the air you have breathed out will collect in the container and by comparing the water level inside with the graduations marked on the container you will obtain an approximate measure of the maximum volume of air which your lungs can breathe in and out. This is called your *vital capacity*.

(e) If you intend to repeat the experiment, you must remove enough water from the basin to restore the level to the 50 mm mark.

Tidal volume

(f) When you have measured your vital capacity, leave the air in the container and the bowl nearly full of water but push the rubber tubing further through the neck of the container until it is above the water level inside, position B (Fig. 1).

(g) Blow through the tube just enough to clear out any water trapped in it and raise or lower the container until the water level inside and outside is the same and at a convenient mark, e.g. the 4 or 3.5 litre mark.

(h) Place the end of the rubber tubing in your mouth and breathe in and out a few times as normally as possible, holding the container and allowing it to move up and down in the bowl so that the water levels inside and outside remain the same.

(i) By watching the graduations on the side, you will be able to see approximately how much air you exchange when you breathe normally. This is called the *tidal volume*.

(j) If a second person is to use the apparatus, the rubber tubing should be rinsed in disinfectant and washed under the tap.

Exhale through the tubing into the container

Experiment 4.   Discussion

1 What results did you get for **(a)** vital capacity, **(b)** tidal volume?
2 Why do you think the vital capacity is so much greater than the tidal volume, i.e. if you need to exchange only a few hundred cubic centimetres of air in breathing, what is the advantage of having a lung capacity of several litres?

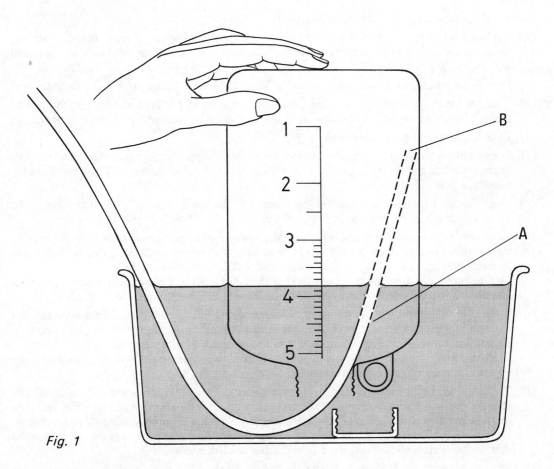

Fig. 1

Experiment 5. Analysis of air

Sodium hydroxide solution absorbs carbon dioxide; alkaline pyrogallol (pyrogallic acid dissolved in sodium hydroxide solution) absorbs oxygen and carbon dioxide. Both the solutions are caustic (i.e. they dissolve skin and fabrics). If spilt on the skin wash the liquid off at once under the tap with much water until the 'soapy' feeling is gone. If spilt on the clothing, neutralize with *dilute* hydrochloric acid and then wash with water. Keep the solutions in the tray and conduct all operations over the tray.

(a) Copy the table on p. 237 into your notebook.

(b) Put some water in the jar to a depth of about 3 cm and add some coloured liquid. Place the microburette upright in the jar with the bulb in the water. All operations must now be conducted with the bulb in the liquid.

(c) Detach the syringe, fill it with water from the jar and replace it in the three-way junction.

(d) Turn the tap to the horizontal position, check that the screw clip is open and depress the syringe plunger briskly to fill the apparatus with water. Withdraw the plunger to refill the syringe.

(e) Collect a sample of gas as described in (m).

(f) Connect the syringe containing the gas sample to the side arm of the three-way junction, turn the tap upwards and force the gas from the syringe into the microburette until it starts to bubble out of the bulb.

(g) Turn the tap horizontal and depress the plunger of the top syringe so that all the gas in the burette is driven down into the bulb and any excess escapes from the hole in the end.

(h) Keep the bulb in the water and withdraw the plunger to bring the air back into the stem of the burette. Check that the air column is unbroken. If there are any bubbles or the column is interrupted with water, return it to the bulb and withdraw it again carefully.
Use the screw clip to adjust the lower level of the air column to a convenient point on the scale and then measure the length of the air column. Record this in column *A* in your table.

(i) If there is sufficient time, return the air column to the bulb, draw it back into the tube and measure it again to ensure consistent readings or to obtain an average reading. (An *increase* in the column between readings could be due to the inclusion of water droplets in the column.)
If any air enters or escapes from the bulb after this stage, the results will be invalid and the experiment must be started again from (b) after washing the burette.

(j) Push most of the air back into the bulb and transfer the microburette to the sodium hydroxide solution. Draw the air into the tube and return it slowly to the bulb three or four times and then measure the air column again. Record this length in column *B* in your table. Measure the air column a second and a third time after returning the air to the bulb to ensure that all the carbon dioxide has been absorbed and there is no further reduction in volume.

(k) Return the air to the bulb and transfer the burette to the alkaline pyrogallol. As before, move the gas between the bulb and the tube several times and take two or three measurements until there is no further reduction in volume. Record the length in column *C* in your table.

(l) Use the results to calculate the percentage oxygen and carbon dioxide in the gas sample. Wash the burette thoroughly before analysing the next sample.

(m) *Collecting the sample.* The sample is collected in a 2 cm^3 syringe and the method will depend on the source of the sample. For exhaled air, expel most of the air from your lungs and blow the last fraction into the nozzle of a *clean* syringe while withdrawing the plunger. Connect the syringe without delay to the side arm of the three-way tap.

	A Length (air)	B Length (CO_2 absorbed)	C Length (O_2 absorbed)
Average			
Reduction due to absorption of CO_2 or O_2		A minus B	B minus C

Calculation

$$\frac{\text{1st reduction in length } (A - B)}{\text{original length } (A)} \times 100 = \% \ CO_2$$

$$\frac{\text{2nd reduction in length } (B - C)}{\text{original length } (A)} \times 100 = \% \ O_2$$

Experiment 5.   Discussion

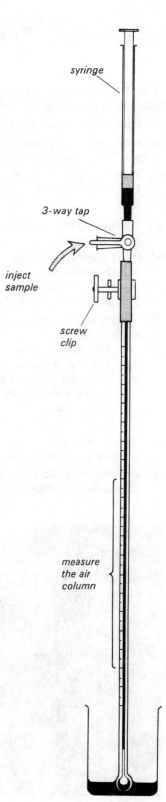

syringe

3-way tap

inject sample

screw clip

measure the air column

1 What differences were there in the composition of exhaled and atmospheric air, according to your results?

2 If alkaline pyrogallol absorbs both carbon dioxide and oxygen, how can it be claimed that only oxygen is absorbed by it in the last part of the experiment?

3 Apart from the absorption of carbon dioxide and oxygen what changes in conditions might affect the length of an enclosed column of air?
How likely is it that such changes might have affected your results?

4 (a) What result did you obtain for the percentage carbon dioxide in atmospheric air?

(b) Consecutive readings of the same volume of air are likely to differ by as much as 0.5 mm. What percentage error will this give with a 10 cm column of air?

(c) The percentage volume of carbon dixoide in atmospheric air is about 0.04%. Why can this not be measured by the burette in this experiment?

5 The burette measures only the *length* of the air column and not its volume. Why is it nevertheless reasonable to use the results to calculate the percentage *volumes* of oxygen and carbon dioxide?

Experiment 6.   Gaseous exchange during respiration

Study Fig. 1 for a few seconds to see how the apparatus is designed, but do not assemble it yet. Soda-lime is a chemical which absorbs carbon dioxide.

(a) Fill a beaker or jar two thirds full with cold water and place it on a tripod and gauze.

(b) Label two boiling tubes A and B; place the living material in tube A and an equal quantity of non-living or dead material in tube B.

(c) Push a plug of cotton wool into each tube. If the living material is blowfly larvae (maggots), leave a gap between the larvae and the cotton wool or they will wriggle through into the soda-lime.

(d) Put equal weights (5 g) of soda-lime on top of the cotton wool in each tube and place both tubes in the water bath.

(e) If the manometers do not already contain liquid, use a dropping pipette to half fill the reservoir of each manometer with a coloured liquid. Air bubbles can be removed from the capillary by blowing and sucking down the long arm of the manometer.
Fit the manometers into the rubber tubing at T.

(f) Check that the taps are open (horizontal), and fit the bungs securely into the boiling tubes. Replace the boiling tubes in the water bath and check for leaks as follows:
   (i) Fit a syringe, with plunger pushed in, to one of the three-way taps.
  (ii) Carefully withdraw the syringe plunger to make the liquid rise up the manometer.
 (iii) Close the tap (downwards) and watch the level of liquid in the manometer. If it falls, there is a leak and the bung, manometer connection and tap connection should be checked, in that order.
 (iv) Repeat the operation with the second tube.

(g) Remove the syringe, open both taps (horizontal), move the rubber marker to indicate the level of liquid in the manometer and close both taps (downwards).

(h) Observe the liquid levels in the manometers for 5-15 minutes. If the level rises to the top of the manometer or falls to the bottom, open the tap to restore the original level.

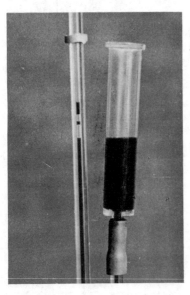

Remove air bubbles from the capillary

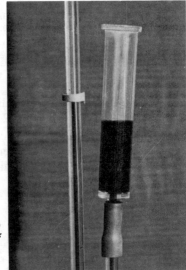

Move the marker to indicate the level of the liquid

Experiment 6.　Discussion

1　What change in the liquid level did you observe in the two tubes?
2　What does this change tell you about the volume of air in each boiling tube?
3　What are the principal gases in the atmosphere and what is the percentage volume of each?
4　What is the function of the soda-lime in this experiment? What might happen if it were not included?
5　Which atmospheric gases could be responsible for the change in the volume of air in the boiling tube?
6　Say why you regard your results as reasonable evidence that respiration is taking place in the organisms under investigation.
7　Why were the two tubes placed in a water bath?
8　Tube B is a 'control' to the experiment. What is its purpose?

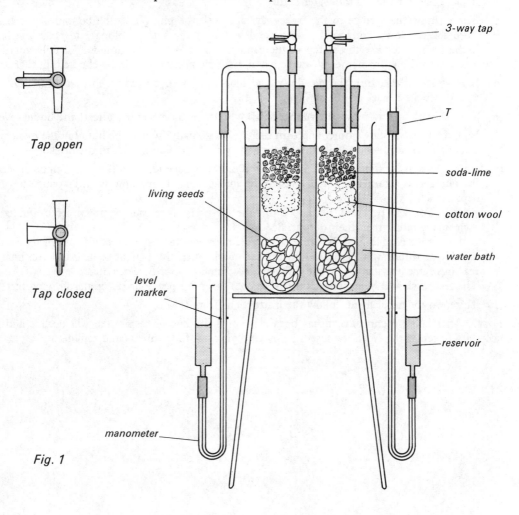

Fig. 1

239

Experiment 7. Measuring the uptake of oxygen

Study Fig. 2 to form a clear idea of the design of the apparatus but do not assemble it yet. Note also the diagrams showing the function of the three-way tap.

(a) Fill a beaker or jar two thirds full with cold water and place it on a tripod and gauze.

(b) Place a known weight of living material in the boiling tube and then push a plug of cotton wool into the tube. If the living material is blowfly larvae, leave a gap between the larvae and the cotton wool or they will wriggle into the soda-lime.

(c) Put about 5 g soda-lime on top of the cotton wool and place the tube in the water bath.

(d) If the manometer does not already contain liquid, use a dropping pipette to half fill the reservoir of the manometer with coloured liquid. Air bubbles should be removed by gently blowing and sucking down the long arm of the manometer.

(e) Check that the tap is pointing upwards and fit the bung and manometer securely into the tube. Replace the test-tube in the water bath.

(f) Before fitting the syringe to the three-way tap, push the plunger to the bottom of the barrel. Turn the tap to its horizontal position and withdraw the syringe plunger to make the liquid in the manometer rise to the top of the capillary tube. Watch the liquid. If it falls, there is a leak and the bung, manometer connection and tap connection should be checked, in that order.

(g) If there is no leak, turn the tap down and move the syringe plunger so that its base is level with the highest mark on the syringe.

(h) Turn the tap up and move the rubber marker on the manometer to indicate the liquid level.

(i) Note the time, or start a timer and turn the tap horizontally. Watch the liquid in the manometer and at frequent intervals depress the syringe to return the liquid level to the marker.

(j) After a convenient time interval, e.g. 1, 5 or 10 minutes depending on the rate of uptake, depress the plunger to return the liquid level to the marker and record how far the syringe has been depressed.
(NOTE.   You will probably need to count the divisions since most syringes are calibrated for their intake rather than their output. See Fig. 1.)
If the syringe plunger is fully depressed before your selected time interval is complete, turn the tap down, withdraw the plunger to the top mark, turn the tap horizontal and continue the readings remembering to add the full syringe volume to your final reading.
The results should be expressed in cm^3 oxygen taken up per hour per gram of living material.

(k) To repeat the experiment, follow the instructions from **(g)**.

NOTE.   After about 6 cm^3 oxygen has been consumed the rate of respiration will decline and it is necessary to renew the air in the test-tube by emptying out the contents and refilling it.

Experiment 7. Discussion

1 **(a)** Why was soda-lime included?
   (b) What might have happened if soda-lime had been omitted?
2 What was the point of keeping the tube immersed in water?
3 On what grounds can one assume that it is oxygen rather than any other atmospheric gas which is taken up?
4 Why should the results be expressed as '*per gram*' of living material?
5 If you used the apparatus to compare the rates of respiration in two or more samples of living organisms, suggest reasons for any observed differences.

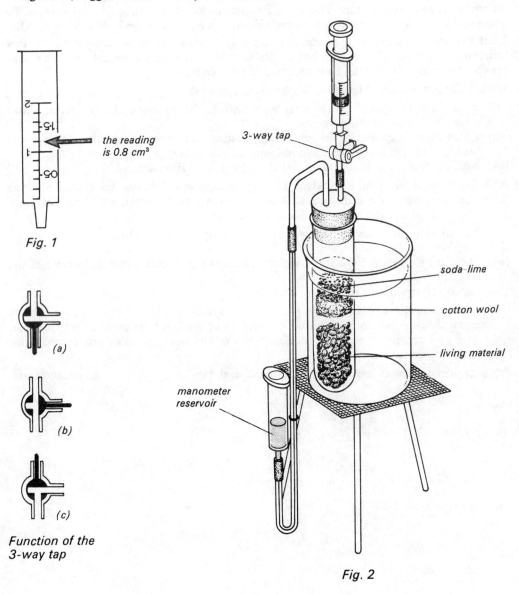

the reading is 0.8 cm³

Fig. 1

(a)

(b)

(c)

Function of the 3-way tap

3-way tap

soda-lime

cotton wool

living material

manometer reservoir

Fig. 2

Experiment 8.   The effect of temperature on respiration

The apparatus and method are similar to those in experiment 6 (p. 238) but if the water bath is to be heated with a Bunsen burner, a beaker or tin can and not a glass jar must be used.

It is not practicable to use the same material and apparatus for measurement of oxygen uptake at several different temperatures so it is necessary to conduct parallel experiments using the same weights of material subjected to different temperatures. Thus it is best for each group to measure the oxygen consumption at one temperature and then pool the results.

The temperatures to try are 20, 30, 40, 50 °C, or if there are 5 groups, 15, 25, 35, 45, 55 °C. With more than five groups, intervals of 5° from 20° to 50° or 15° to 55° may be feasible.

(a)   Having selected a temperature, prepare a water bath by filling a tall beaker or tin can two thirds full with water at room temperature. Follow the instructions for experiment 6 (p. 238) as far as **(f)**. Make sure that all groups are using the same weights of material and check for leaks as in **6 (f)**.

(b)   Check that the taps are open (upwards) and then heat or add ice to the water until the required temperature is reached. Leave both tubes, with their taps open, in the water bath at this temperature for at least 10 minutes before starting to take readings.

(c)   While waiting, copy the tables on p. 243 into your notebook.

(d)   **(i)** Ensure that the syringe plungers are withdrawn to the top mark and the marker shows the level of liquid in the manometer.

   (ii) Check the temperature of the water bath, heating or cooling it as needed. Do not heat the water bath in the period when readings are being taken.

   (iii) Note the time, or start the timer, and close the taps (horizontal).

(e)   After five minutes (or more, if the change in level is very small) depress the syringe plungers to restore the levels in the manometers. Read the volumes on the syringes to find how much air has been added and record these volumes in your table.

(f)   **(i)** Turn the taps down and withdraw the syringe plunger to the top mark.

   (ii) Turn the taps up.

   (iii) Check the temperature and if it is different from the first reading record the mean of the two.

   (iv) Repeat the experiment twice more from **(d)**.

(g)   Subtract the control readings (B) from the experiment readings (A) and express the results as cm^3 oxygen used up per hour. Calculate the average of the three results and also the mean temperature for the three experiments. Collect the results from the other groups and enter them in the second table.

(h)   Plot a graph of the oxygen consumption against temperature with the latter on the horizontal axis.

Weight or number .... Organisms			
Temperature			
Tube A reading			
Tube B reading			
Net uptake (A — B)			
Uptake of O_2 in cm^3 per hour			
Average uptake of O_2 at ...... °C ...... cm^3/h			

Class results										
Temperature										
Oxygen uptake										

Experiment 8.　Discussion

1　Why was it necessary to subtract readings B from readings A?

2　How much difference was there between your three readings? Suggest reasons for the differences, if any.

3　From the graph, comment on any trend in the oxygen uptake with increasing temperature. Suggest reasons why changes in temperature should affect the rate of oxygen uptake.

4　If your results include temperatures above 50 °C, comment on and suggest reasons for the results. If you do not have results for the high temperatures, say whether you think that the trend in results for the lower temperatures would continue with temperatures of 50, 60 and 70 °C. Give reasons.

Experiment 9.   Oxygen uptake in blowfly larvae

The apparatus and method are those described in experiment 6 but in this case equal weights of living blowfly larvae are used in each of the tubes A and B.

(a) Copy the table below into your notebook.

(b) Set up the experiment, following the instructions for experiment 6 (p. 238) as far as (f) using equal weights (10 g) of living blowfly larvae in each tube A and B.
DO NOT INSERT THE BUNGS UNTIL THE WATER BATH AND BOTH TUBES ARE READY.
Check for leaks as described in 6 (f).

(c) Check that
(i) the taps are open (upwards)
(ii) the syringe plungers are withdrawn to the top mark
(iii) the markers indicate the liquid level in the manometer.

(d) You will need to record the time on the clock as well as the time interval for each reading. Note the time and close the taps (horizontal) for five minutes.

(e) During the five minutes, depress the syringe plunger as necessary to prevent the liquid rising above the capillary tube and at the end of the five minutes depress the syringe to return the liquid level to the mark.

(f) Open the taps (upwards). Note and record the volume of air added from each syringe.

(g) Detach the syringe, return the plungers to the top mark and replace them in the three-way tap. Note the time and close the taps (horizontal) for a further five minutes and again measure the volume of air added.
Open the taps (upwards).

(h) Repeat the operations in (g) twice more to obtain a total of four readings over a period of about 20 minutes.

(i) After this period of about 20 minutes, note whether the maggots are actively wriggling or fairly quiescent and then transfer the maggots, soda-lime and cotton wool from tube A to a fresh tube and connect it to the manometer again.

(j) Without opening tube B, shake it about to disturb the larvae and make them wriggle. Replace both tubes in the water bath, leave for one minute and then following the instructions from (c), take four more five-minute readings.

(k) Calculate the oxygen uptake in cm^3/h and plot a graph of oxygen uptake against time (i.e. the time elapsed since the start of the experiment). The time is on the horizontal axis.

	Before disturbing				*After disturbing*			
Time								
Time interval								
Reading, tube A								
Reading, tube B								
Uptake of O_2 in cm^3/h in A								
Uptake of O_2 in cm^3/h in B								

Experiment 9.   Discussion

1 Was there any consistent change in oxygen uptake in the first four readings?

2 Suggest reasons for any changes if these were discernible.

3 What was the effect on oxygen uptake of (a) transferring the maggots to a new tube and (b) shaking them up without opening the tube? Suggest reasons for any difference of results between tubes A and B.

4 Were any of the second series of readings higher than any of the first four? If so, what explanations can you suggest.

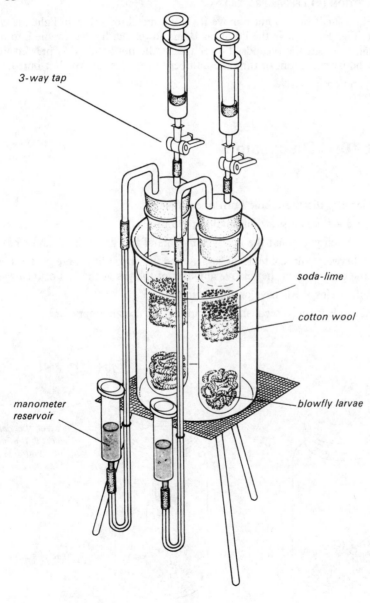

3-way tap

soda-lime

cotton wool

manometer reservoir

blowfly larvae

Experiment 10.   The gas used up in respiration

You are provided with two screw-top jars and some germinating seeds.

(a) Label the jars A and B. Add your initials and the date.

(b) Place some cotton wool in the bottom of each jar and add enough water to just moisten it.

(c) Count an exact number of germinating seeds into jar A and the same number of dead seeds into jar B.

(d) Close each jar, screwing the lid down as tightly as possible. Leave the jars for 24 hours or more.

READ ALL OF INSTRUCTION **(e)** BEFORE PROCEEDING.

(e) Loosen the lid on jar B but do not remove it. Use a match or splint to light the candle on the special holder (Fig. 1). Remove the lid of jar B and lower the lighted candle into it, noting the time or starting to count the seconds. Lower the candle until the lid is pressed firmly against the mouth of the jar and count or time the number of seconds for which it burns.

(f) Repeat this operation for jar A.

Experiment 10.   Discussion

1 For how many seconds did the candle burn in each jar?

2 Which atmospheric gas is necessary for anything to burn?

3 What do the results tell you about the relative amounts of this gas in jar A and jar B?

4 What difference between the contents of jars A and B when they were first set up could be responsible for the difference in the relative amounts of the gas at the end of the experiment?

5 What effect did the living seeds have on the air in the jar?

6 Why was it better to have dead seeds in jar B rather than use an empty jar?

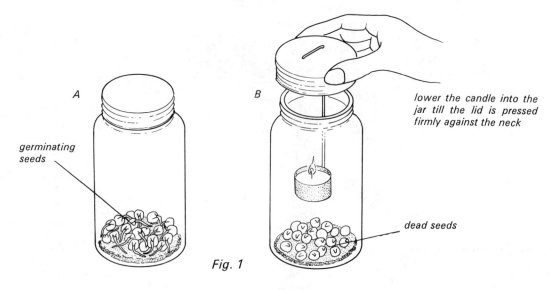

A

B

germinating seeds

lower the candle into the jar till the lid is pressed firmly against the neck

dead seeds

Fig. 1

Experiment 11. The gaseous product of respiration

(a) Label two boiling tubes A and B and place a known weight or number of living organisms in A. Place an equal quantity of dead or non-living material in B.
Cover the mouth of each tube with a piece of aluminium foil. Press firmly into place round the rim of the tube.
Leave the tubes for at least ten minutes.

(b) Use a graduated pipette to place 2 cm³ lime water in each of two clean test-tubes and label these A and B.

(c) You are provided with a syringe attached to a glass delivery tube drawn out to a fine point. Check that the syringe plunger is pushed fully down in the barrel.

(d) After the experiment has been running for ten minutes or more, push the delivery tube, still attached to the syringe, through the foil cap of tube B. Keeping the point of the delivery tube close to the side of the boiling tube to avoid damaging the organisms, insert the delivery tube into the midst of the organisms. Withdraw the syringe plunger to fill the syringe with air from the boiling tube.

(e) Place the syringe and delivery tube in test-tube B containing lime water so that the delivery tube is below the level of the liquid and gradually depress the plunger so that the air from the syringe bubbles slowly through the lime water. When all the air has been expelled from the syringe, remove the delivery tube, close the mouth of the test-tube with your thumb (or a cork) and shake the tube thoroughly for a few seconds.

(f) Repeat the operation from (d) for boiling tube A, using test-tube A of clear lime water.

(g) Compare the appearance of lime water in the two tubes.

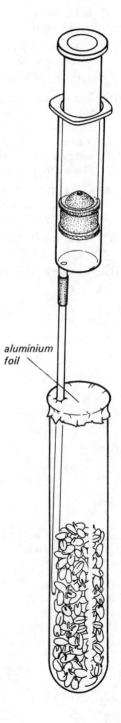

aluminium foil

Experiment 11. Discussion

1 What difference did you observe between the two tubes containing lime water at the end of the experiment?

2 What gas affects lime water in this way?

3 What gas is given out by the living organisms in your experiment?

4 How do you know that this gas comes from the living organisms and not from the air already in the tube?

5 From your results can you say that all living organisms produce this gas?

6 If you used dead organisms in tube B suggest a reason for having them rather than an empty tube.

Experiment 12.   Anaerobic respiration in yeast

(a) Label two boiling tubes A and B and place 1 g dried yeast in A. In B place 1 g yeast which has been heated to 100 °C for 15 minutes.

(b) Pour about 30 mm (depth) lime water into each of two clean test-tubes. Half fill a beaker or jar with warm water at about 40 °C.

READ ALL OF INSTRUCTION **(c)** BEFORE PROCEEDING.

(c) Use a graduated pipette to place 10 cm³ glucose solution in each of tubes A and B. The glucose solution has been made with boiled water. As you release this solution into the tubes keep the tip of the pipette in contact with the inside of the tube so that the liquid runs down the tube without splashing and introducing air. Swirl the tubes gently to mix the yeast and glucose solution but do not shake them and so introduce air.

Use a dropping pipette to cover the surface of the solution in each tube with a thin layer of oil.

(d) Check that the taps are open (pointing upwards) and then fit the rubber bungs securely into each tube. Place both tubes in the water bath so that their delivery tubes are dipping into the lime water in the two test-tubes.

(e) Leave the boiling tubes to acquire the temperature of the water bath for a minute, then close the taps (pointing downwards).

(f) Observe any changes in the lime water and in the tubes with the yeast and glucose solution.

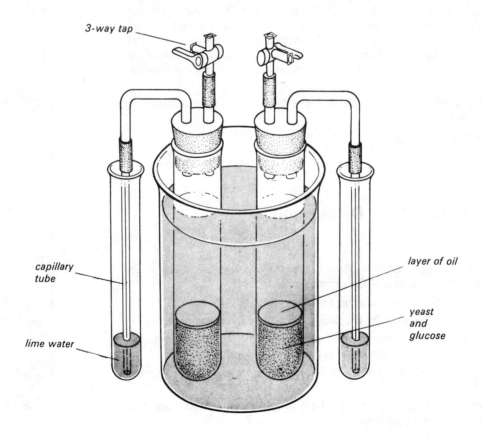

3-way tap

capillary tube

layer of oil

yeast and glucose

lime water

Experiment 12. Discussion

1 What changes did you observe in the lime water and in the tubes containing the yeast and glucose?

2 How do these observations support the idea that respiration is taking place in the yeast?

3 Suppose it is argued that the results are due not to respiration but to a simple chemical reaction between the yeast and glucose, in much the same way as hydrochloric acid and calcium carbonate react, what evidence do you have from the experiment to refute this argument?

4 It is further argued that by bubbling air through lime water, the same results could be achieved. What control experiment could you carry out to show that this was not the case in your experiment?

5 Assuming that the results *are* due to respiration in the yeast, how was the experiment designed to show that it was *anaerobic* respiration (i.e. respiration in the absence of oxygen)?

6 What was the role of the glucose solution in this experiment?

Run the liquid carefully down the side of the tube

Experiment 13. The effect of temperature on fermentation rate

You are provided with a boiling tube containing a suspension of yeast in a solution of glucose. DO NOT SHAKE THE TUBE at any stage of the experiment.

(a) Set up a clamp stand, clamp, tripod, gauze and water bath as shown in Fig. 1. Clamp the boiling tube securely in the water bath.

(b) Fit the manometer and syringe assembly as shown in the diagram ensuring
   (i) that the bung makes an air-tight seal in the mouth of the tube and
   (ii) the tap is open (turned upwards).

(c) Use a dropping pipette to fill the reservoir of the manometer about three quarters full of coloured liquid. If the liquid does not flow into the capillary, close the tap (horizontal) and raise the syringe plunger. Air bubbles in the column can be expelled by using the syringe to push the liquid and trapped air back to the reservoir.

(d) Turn the tap downwards, push the syringe plunger to the bottom of the barrel and open the tap (vertically upwards). Move the marker on the capillary tube to indicate the level of liquid and then copy the table on p. 251 into your notebook.

(e) Record the temperature of the water bath, note the time and close the tap (horizontal). Leave the tap closed for one minute and keep withdrawing the syringe plunger to bring the liquid back to the marker. After one minute bring the liquid level exactly back to the mark, open the tap (upwards) and record in your table the volume indicated on the syringe.
This figure represents the volume of carbon dioxide given out by the yeast in one minute.

(f) Turn the tap downwards, push the syringe plunger to the bottom and open the tap (upwards).

(g) Take two more readings of the volume of gas produced in one minute and calculate the average of the three.

(h) Check that the tap is open (upwards) and use a small Bunsen flame to raise the temperature of the water bath by about 5 °C. Leave the apparatus for two minutes after heating has stopped. Record the new temperature and take three more one-minute readings as before but this time record the temperature at the end of the three readings to obtain the mean temperature during the experiment.

(i) Repeat the operation from **(g)** raising the temperature about 5 °C each time until you reach 45 or 50 °C.
At temperatures between 30-40 °C the three readings should be made in rapid succession to avoid volume changes due to cooling; at 40-50 °C you should add 0.1 cm^3 to all readings to compensate for the cooling effect.

(j) Plot a graph of carbon dioxide output (vertical axis) against temperature.

Experiment 13.   Discussion

1   In general terms what was the effect of rising temperature on the rate of carbon dioxide production?

2   Was the change in rate consistent over the whole of the temperature range which you tried? If not, suggest an explanation for any deviation.

3   As the water bath cooled down during each series of three readings, there would be a contraction in the volume of gas. How could this affect the accuracy of your readings? Suggest a control which could be employed to correct any errors of this kind.

4   The solubility of gases in water decreases with increasing temperature. How might this property affect your results?

5   What factors, apart from temperature, might be limiting the rate of carbon dioxide production from yeast cells?

Temperature	Mean temperature	Volume	Mean volume
1st			
final			
1st			
final			
1st			
final			
1st			
final			

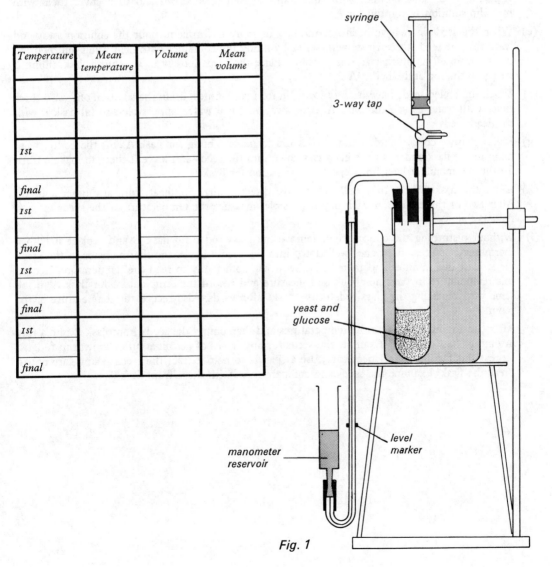

Fig. 1

251

Experiment 14. Energy release during respiration

(a) You are provided with a quantity of wheat grains which have been soaked for 24 hours. Divide these into two approximately equal batches A and B.

(b) Place the wheat from batch B in a large beaker or tin can, cover the grains with tap water and heat them to boiling point over a Bunsen flame. Allow the water to boil for five minutes to kill all the seeds. During this period, continue with instruction **(c)**.

(c) Label a suitable container with the letter A and place the other batch of seeds in it. Cover the wheat with a 1% solution of formalin and leave it for five minutes or more. This is to kill fungi and bacteria on the surface of the grains.

(d) After B has boiled for five minutes, turn out the flame and add cold water to the container to cool down the contents. As soon as the container is cool enough to handle, drain off the water, rinse the seeds several times with cold water to cool them down and then cover them with formalin solution as you did for A.

(e) When the seeds in A have been five minutes or more in formalin, pour the solution away and wash the live seeds three times with water by filling the container with cold water and pouring it off. Drain off as much water as possible. Place the seeds loosely in a vacuum flask, filling it nearly to the top and label it 'A'.

(f) When the boiled seeds have been in formalin for five minutes, drain the solution off, wash three times with water as you did with A, pour away as much water as possible and fill flask B with the dead seeds.

(g) Roll a strip of cotton wool round a glass rod to make a bung for flask A. Fit the cotton wool bung in the flask and push the glass rod down into the seeds and leave it there to make a track for the thermometer later on. Repeat the operation for flask B.

(h) Leave the flasks where they will not receive direct sunlight or heat from any other sources. After two or three days, or each day if possible, measure the temperature of the seeds in each flask as follows:

(i) Without disturbing the cotton wool, remove the glass rod from flask B and replace it with a thermometer so that it reaches well down into the seeds. Leave the thermometer in the seeds for at least two minutes and then withdraw it just sufficiently to read the temperature. Record this temperature in your notebook and measure and record the temperature for flask A in the same way, replacing the glass rods in the flasks afterwards. Also record the temperature of the room.

(j) After two to four days, note the temperatures and then empty the seeds from flask B into a tray or onto a sheet of paper. Examine the seeds for any traces of germination or growth of fungus. Repeat this for flask A. If necessary, use a spatula to help extract the seeds taking care not to break the flask. Examine the seeds and compare them with those from flask A.

Experiment 14. Discussion

1 Why was it necessary to use insulated (vacuum) flasks for the experiment?

2 What difference was there between room temperature and the temperature in the flasks after two to three days?

3 What evidence is there from this experiment that any heat produced is a result of living activity in the seeds?

4 Is there any evidence to confirm that any heat produced is a result of respiration in the seeds?

5 If there had been signs of a vigorous growth of fungi or indications of bacterial decay, how would this have affected your interpretation of the results?

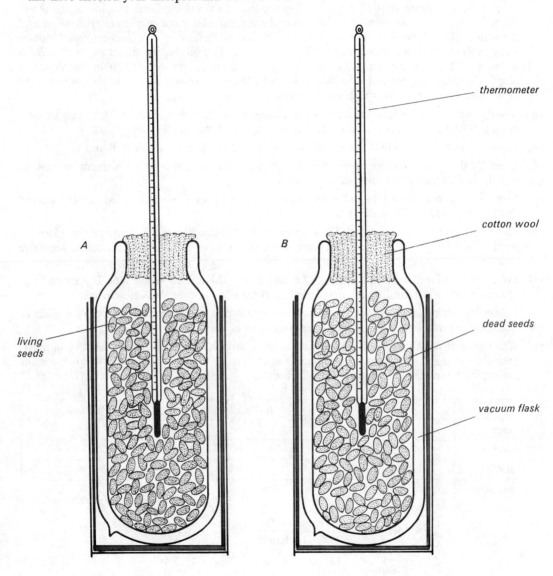

thermometer
cotton wool
A
B
living seeds
dead seeds
vacuum flask

Experiment 15.　Energy for muscle contraction

In this experiment, the substance under investigation is adenosine triphosphate (ATP). This substance occurs in all living cells. The tissue used in this experiment is muscle. The contraction of the muscle is evidence that energy is being set free. Ringer's solution is a solution of salts in water of about the same strength as tissue fluid.

(a) Copy the table given below into your notebook.

(b) Select three clean microscope slides and label them A, B and C by writing close to one end of the slide.

(c) Place a strip of muscle (meat) on a wooden board and use a scalpel and forceps to cut three very thin strips. The strips must be cut parallel to the line of the muscle fibres and not across them. A good method is to smooth, straighten and slightly flatten a piece of muscle with the flat of the scalpel blade. Then cut the muscle by rocking the scalpel, i.e. with a 'rocking chair' motion while pushing it forward slowly along the strip (see Fig. 1). Pull the fine strip away with forceps as you cut. Take the fine strip, flatten it on the board and, if possible, cut it in two down its length. Keep cutting the strips in half down their length to obtain the finest possible filament of muscle, 1 mm or less in width and between 20 and 40 mm long.

(d) Cut the ends of the muscle filament square, place it on slide A and with a dropping pipette, put a few drops of Ringer's solution on the slide to cover the muscle fibres.

(e) Repeat this operation to obtain and mount similar strips of muscle on slides B and C.

(f) Carefully tilt slide A to drain off most of the solution. Straighten the muscle filament by pushing it with the forceps but do not pull it or stretch it.

(g) Hold the slide over the millimetre scale of a ruler and measure the length of the muscle. Record this length in column A of the table.

(h) Collect the syringe containing ATP solution and carefully depress the plunger to place no more than 0.05 cm³ (Fig. 2) ATP evenly along the muscle. Leave this slide for at least five minutes.

(i) Collect the syringe containing boiled ATP and repeat the operation for slide B starting from instruction (f) and recording the first length of the muscle in column B of your table.

(j) Collect the syringe containing glucose solution and repeat the operation from (f) for slide C, recording the first length of the muscle in column C of your table.

(k) When each of the slides has been exposed to its appropriate solution for at least five minutes, measure the lengths of the muscle filaments again, record this in your table and work out the amount of contraction and percentage contraction.

	A (ATP)	B (Boiled ATP)	C (Glucose)
1st length			
2nd length			
Decrease			
% contraction			

$$\% \text{ contraction} = \frac{(\text{1st length} - \text{2nd length})}{\text{1st length}} \times 100$$

Experiment 15.   Discussion

1 Which of the three solutions caused the greatest percentage contraction?

2 Glucose is generally considered to be the main source of energy for reactions such as muscle contraction (e.g. athletes may eat glucose tablets before strenuous effort). Discuss whether your results support this view.

3 Is there any evidence, one way or the other, to suggest that ATP is acting as an enzyme in promoting muscle contraction?

4 The animal from which the muscle was obtained has probably been dead for many days. Does the fact that muscle fibres will still contract mean that the muscle is still alive?

5 Discuss whether the results of the experiment entitle you to say that ATP causes muscle contraction in living organisms.

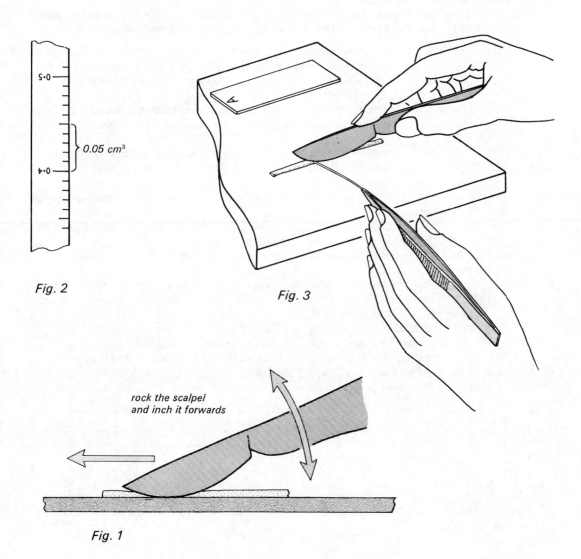

5·0

0.05 cm³

0·4

Fig. 2

Fig. 3

rock the scalpel
and inch it forwards

Fig. 1

Experiment 16.   Changes in weight during germination

(a) You are provided with about 100 soaked wheat grains. Take ten of them, place them in a test-tube with a little water and, using a test-tube holder, boil them over a small Bunsen flame for half a minute. Add cold water to the tube to cool it down, pour off most of the water and cover the seeds with $2\frac{1}{2}\%$ formalin for about ten minutes. Meanwhile, copy the table below into your notebook and continue with instruction **(b)**.

(b) Select ten more grains, weigh them, record their weight (wet weight) and place them in an open container labelled with the letter A, the date and your initials. Place this container in an oven at 50 °C, or on a radiator, or simply leave it open in the laboratory (air drying).

(c) Place the remaining soaked grains in a suitable container with a lid and mark it with the date and your initials.

(d) Pour off the formalin from the boiled seeds and place them in a container labelled 'control' and add the date and your initials. Close the container.

(e) The germinating seeds are now left for eight days or more, taking samples, drying and weighing them at two-day intervals as follows (Day 0 is the day the experiment was started):

DAY 2   **(i)** Select ten germinating seedlings, weigh them, record their (fresh) weight in your table. Place them in an open container labelled with the letter B, the date and your initials. Place this container in the oven, or on a radiator, or leave it in the laboratory to air-dry the seedlings as before.

(ii) If the seeds in A were in the oven or on a radiator, weigh them and record their (dry) weight in your table. If the seeds in A are being air-dried, do not weigh them till Day 4.

(iii) Check that the germinating seeds are not drying out. Add water if necessary.

DAY 4   **(i)** Repeat the operations described for Day 2 placing the ten freshly selected and weighed seedlings in a container labelled C.

(ii) Weigh the dried seedlings in B if they were oven- or radiator-dried. Otherwise weigh the dried seeds in A.

(iii) Remove the lid from the remaining growing seedlings, leave them in the light and check each day that their roots are moist.

(iv) If you are air-drying the seedlings, remove the lid from the 'control' container and leave it open till Day 8.

DAY 6   **(i)** Continue the operations as before selecting and weighing ten more seedlings (D).

(ii) If you are using an oven or a radiator, remove the lid from the 'control' container and place it in the oven or on the radiator.

(iii) Weigh the dried seedlings in C (or B if air-dried).

DAY 8   **(i)** Select ten more seedlings, weigh them and place them in container E.

(ii) Weigh the dried seedlings in D (or C if air-dried) and also the control seeds.

DAY 10 Weigh the dried seedlings in E (or D if air-dried).

DAY 12 Weigh the dried seedlings in E (if air-dried).

	Control	A (Day 0)	B (Day 2)	C (Day 4)	D (Day 6)	E (Day 8)
Fresh weight						
Dry weight						

Plot a graph of the fresh and dry weights against time and discuss the possible reasons for the changes in fresh and dry weight.

9 Human Senses

Introduction

The experiments in this section mostly depend on subjective experience; that is, it will be necessary for you to decide whether you see or feel or taste something. This often necessitates having two people to conduct the experiment, a subject on whom the tests are performed and an experimenter who administers the tests and records the results.

The aim of the experimenter is to avoid giving the subject any clues about his sensations which are additional to the stimulus being tested, e.g. he will apply touch stimuli at irregular intervals so that the subject does not learn from the rhythm when to expect a stimulus.

There are very wide variations in sensitivity to all kinds of stimuli. Some people have a limited sense of smell, others will not agree about taste or colour and not everyone will experience the visual results intended by experiments 10-17. Very often, the only way to obtain significant results is to collect together as many results as possible from the class and look for a pattern.

Acknowledgments

Several ideas for experiments have arisen from reading *Introduction to Psychology* by Hilgard, Atkinson and Atkinson (Harcourt, Brace, Jovanovich) and from the Open University post experience course, *The Biological Bases of Behaviour*. I am indebted to Professor R. L. Gregory for permission to reproduce the photograph on p. 288 and to Professor T. Cornsweet and Academic Press for permission to use the illustration on p. 279.

Contents

Experiment 1. Sensitivity of the skin to touch

(a) Work in teams of two or three. The person conducting the experiment is the experimenter, the person being experimented upon is the subject. A third person can act as recorder, or the experimenter can do his own recording.

(b) The subject inks the rubber stamp and uses it to mark the areas to be tested on (i) a finger-tip, (ii) the back of the hand and (iii) the inside of the forearm.

(c) The experimenter or recorder draws three grids in his notebook similar to that below. (Alternatively, a printed grid may be provided.)

(d) The subject closes his eyes or looks away so that he cannot see when the experimenter applies the stimulus. The experimenter places a blunt pin in the holder and uses it to test the subject's sensitivity in one of the three areas by touching the pin on to each of the dots in the marked area in turn, working systematically down the rows and *strictly observing the following procedures:*

  (i) rest the pin on the marked area of skin so that the head of the pin is just lifted off the support, keeping the handle horizontal (Fig. 1);

  (ii) try to apply the pin to the *centre* of each mark;

  (iii) leave the pin in contact with the skin for no more than half a second;

  (iv) allow from half a second to one second interval between each touch *but*

  (v) *do not make regular intervals* between each touch or the subject will expect to receive a regular stimulus and this may influence his judgement;

  (vi) do not let any part of your hand or sleeve touch the subject's hand during the experiment.

(e) **Recording the results.** If the subject thinks he can feel the touch of the pin on his skin he says 'yes'. The experimenter or recorder then indicates the responses in the grid in his notebook, leaving blank the squares where the subject makes no response (i.e. he cannot feel the pin). If there is no recorder it is usually convenient for the experimenter to test all five dots in one row and remember which of the stimuli, 1 to 5, evoked a response. These can be recorded before testing the next row.

(f) Each grid should be labelled with the location of the skin tested and the name of the subject. If the number of positive responses is multiplied by 4 the product will give a measure of percentage sensitivity to the pin.

(g) The other two areas are now tested in the same way.

Subject . . . .

Area tested . . . .

Percentage response . . . .

Experiment 1.   Discussion

1 Which of the three areas was (a) most sensitive, (b) least sensitive to touch?
2 From your results, suggest possible explanations for the differing sensitivities of the three areas, in terms of sensory receptors or nerve endings in the skin.

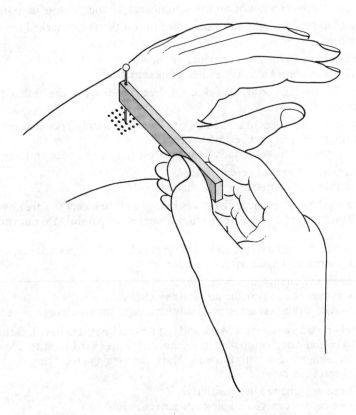

Fig. 1 Allow the pin to rest on the mark for about half a second

Experiment 2.   Sensitivity of the skin to touch: additional experiments

The following experiments are an extension of experiment 1 and it is assumed that the instructions given in that experiment for testing skin sensitivity will be followed exactly as before.

1 **Consistency.** Mark an area of skin which has previously given less than 100 per cent response to a light pin. Test it three times in succession and compare the three records of results.

   (a) Is the pattern of response the same in each, i.e. does stimulation of spots insensitive in one test produce the same lack of response in other tests?

   (b) Is the percentage response the same?

   (c) Do the results throw any doubt on the significance of your results in experiment 1?

2 **Increasing stimulation.** Use light and heavy pins to test the same marked area of skin (other than the finger-tips).

   (a) Does the pattern or percentage response change?

   (b) What are the possible reasons for a change in response?

3 **Different areas.** Try the tests on areas of skin not already tested, e.g. the back of the neck, upper arm or shoulder.

4 **Effect of rubbing.** Mark an area of skin known to have less than 100 per cent response to a light pin. Test it in the usual way and record the results.
   Rub the skin vigorously for 30 seconds and then test it again at once and one minute later.

   (a) Has its sensitivity increased or decreased?

   (b) Suggest possible explanations for any change.

5 **Cooling.** Test a moderately insensitive area as before and then cool the area by holding an ice cube on it for 30 seconds to 1 minute or until it becomes too painful. Do not rub the ice on the skin.
   Gently dab the area dry with a tissue and, if the marks have disappeared, mark the area again. Test the area as before and again after one minute.

   (a) Has there been any change in sensitivity?

   (b) What possible causes can you suggest for any change?

   (c) Can you think of any advantages or disadvantages of such a change of sensitivity?

6 **Effect of alcohol on skin sensitivity.** Mark and test an area known to have less than 100 per cent response. Hold a cotton wool pad soaked in alcohol on the area for 1 minute. Do not rub or dab. Blot up any remaining alcohol with a tissue. Mark the area again if the previous marks have disappeared and test it as before.

   (a) Has there been any change in sensitivity?

   (b) What possible causes can you suggest for any changes?

7 **Discrimination of sensation.** Devise an experiment to see if the subject can distinguish between the sensation of a hair being displaced and the skin being touched by a light pin.

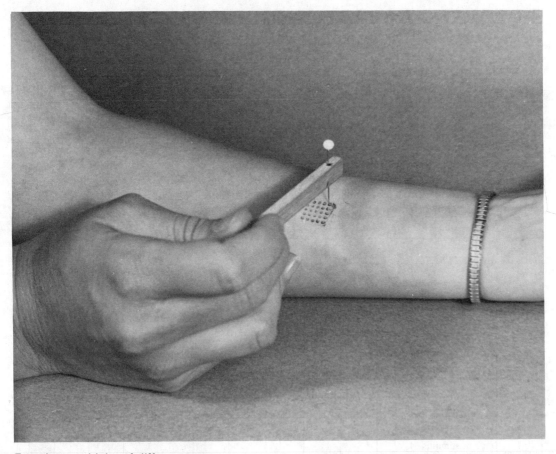

Test the sensitivity of different areas

Try increasing the stimulus

Experiment 3.   Recognition of separate stimuli

(a) There are two or three people in each testing team: the experimenter who applies the stimulus, the subject who receives and reports on the stimulus, and, if a third person is available, a recorder who notes the subject's responses. If there is no third person the experimenter must keep a record of the results.

(b) The object is to determine the minimum distance apart for two simultaneous touch stimuli to be recognized as two sensations rather than one. The subject must not know in advance whether he is to receive one or two stimuli.

(c) The experimenter draws up a table and, without showing the subject, writes down a programme of 5 single and 5 double stimuli, randomly distributed. (A sample programme for the 10 mm gap is shown below.)

Gap between points	AREA TESTED	. . . .										Total correct
*10*mm	type of stimulus	2	1	2	2	1	2	1	2	1	1	
	subject's response											
mm	type of stimulus											
	subject's response											
mm	type of stimulus											
	subject's response											
mm	type of stimulus											
	subject's response											
mm	type of stimulus											
	subject's response											

READ BOTH (d) AND (e) BEFORE STARTING

(d) The subject closes his eyes or turns his head so that he cannot see the stimulus being applied. The experimenter grips the wire 'hair pin' as shown in Fig. 1, adjusting the wires so that the points are exactly 10 mm apart. He then presses the wire carefully on to the skin on the back of the subject's hand so that either one or both wires make a visible indentation in the skin, following his planned programme and observing the procedure given below.

(i) Each stimulus should last for about *half a second*. Do not merely jab at the skin.

(ii) For the double stimulus, both wires must touch the skin *simultaneously*.

(iii) The stimuli are applied at random over the whole area of the back of the hand.

(e) After each stimulus, the subject must say 'one' or 'two' according to how many stimuli he thought he could feel and the experimenter or recorder marks in the table how many stimuli were correctly recognized. After 10 stimuli, the correct responses are added up.

(f) (i) *If the total is 10* the gap between the wire points is reduced to 5 mm and the test repeated. If the repeated test still gives a score of 10, the gap is further reduced and the test repeated until a score of less than 10 is achieved

(ii) *If the score is 9 or less* the test is repeated with the points 15 mm apart. If the score is still less than 10, the test is repeated, increasing the gap between the points by 5 mm each time until a score of 10 is achieved.

(g) Test the front of the wrist and finger-tips in the same way, treating all four finger-tips as the test area.

(h) Retest the area of least sensitivity as follows:

   (i) Set the points to the distance which gave a low score.

   (ii) Follow one of the plans of stimulation previously used but this time, when two stimuli are to be given, separate the stimuli in time by about half a second, i.e. do not apply both points simultaneously as before. Make a note of any difference this method of stimulation has on the successful recognition of a double stimulus.

Experiment 3.   Discussion

1 For each region tested, what were the minimum distances between the points which could still be recognized as separate stimuli?

2 Suggest reasons for the different spatial sensitivity of the regions as shown by your tests. The explanations should consider the distribution and density of touch receptors and their possible connections to nerves.

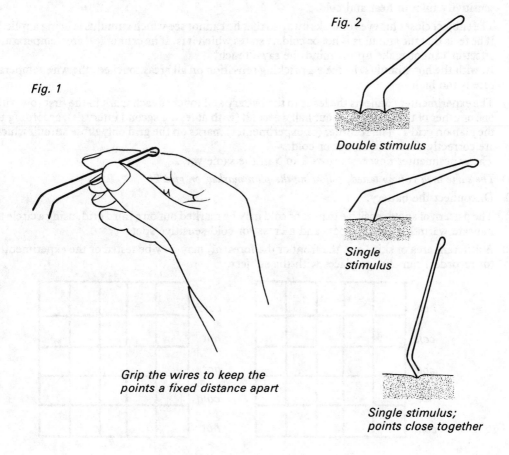

Fig. 2

Double stimulus

Single stimulus

Single stimulus; points close together

Fig. 1

Grip the wires to keep the points a fixed distance apart

Experiment 4. Sensitivity to temperature (1)

DO NOT CONNECT THE BATTERY TO THE APPARATUS UNLESS TESTS ARE BEING CONDUCTED.

(a) The most convenient team consists of three people: an experimenter, who applies the stimulus, the subject who receives and reports on the stimulus and a recorder who notes the subject's responses. If a third person is not available, the experimenter must record the results.

(b) The subject inks the rubber marker and uses it to mark an area of skin on the inside of the upper arm close to the elbow joint.

(c) The recorder or experimenter draws *two* grids in his notebook similar to those in Fig. 1. (Alternatively, a set of printed grids may be provided.)

(d) To conduct the tests, the experimenter will touch each of the marks on the skin with one or other of the two wires on the apparatus in a sequence to be described later. One wire is heated by a small electric current and may feel warm. The other is a copper wire which may feel cold when it conducts heat away from the skin.

(e) The subject must not know in advance which of the stimuli he is to receive. To achieve this, the experimenter, without showing the subject, draws up a plan of the stimuli, so that each row of marks receives either a hot or a cold stimulus, and he indicates this on the first grid as shown in Fig. 1. The second grid is planned so that the rows previously given hot stimuli are given cold stimuli and vice versa. In this way, all the spots marked on the skin are tested for their sensitivity to both heat and cold.

(f) The subject closes his eyes or looks away so that he cannot see which stimulus is being applied. If he feels that the stimulus is hot or cold he states which it is. If he cannot feel any temperature effect or cannot make up his mind, he says 'touch'.
If, with the hot stimulus, he feels a pricking sensation on all areas touched, the wire temperature is too high.

(g) The experimenter connects the leads to the battery and touches each mark in the first row with one or other of the wires for about half a second (with at least 1 second intervals), according to the plan on grid 1. The recorder (or experimenter) marks on the grid only those stimuli which are correctly described as hot or cold.
The experimenter now tests rows 2 to 5 in the same way.

(h) *The same area is now tested, following the plan marked on grid 2.*

(i) Disconnect the battery.

(j) The pattern of spots sensitive to heat or cold may be marked out on a third grid, using a circle to indicate warmth-sensitive spots and a cross for cold-sensitive spots.

(k) A different area of skin, e.g. the front of the forearm, may now be tested or the experimenter (or recorder) can change places with the subject.

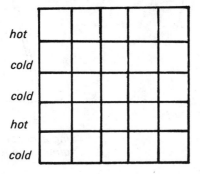
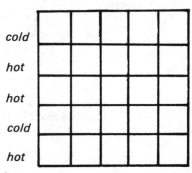

Fig. 1

266

Experiment 4.   Discussion

1 How many of the spots in each area tested seemed quite insensitive to temperature?
2 Suggest possible reasons why certain areas of the skin cannot detect temperature changes in this experiment.
3 Were the spots which responded to heat the same as those which responded to cold?
4 Did any of the hot stimuli evoke a response of 'cold' or vice versa?
   Suggest reasons for such errors of judgement.

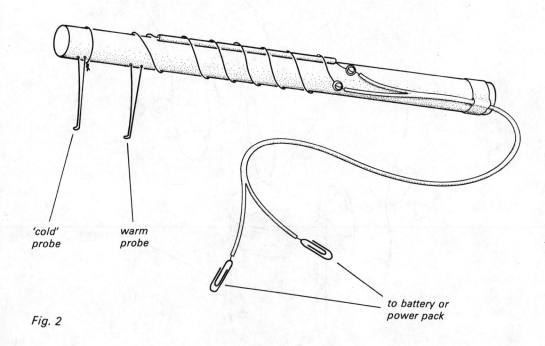

'cold' probe

warm probe

to battery or power pack

Fig. 2

Expermiment 5.   Sensitivity to temperature (2)

(a) Collect three jars or beakers of about the same size. Fill one with cold water (10-15 °C), one with hot water (40-50 °C) and the third with warm water (about 25 °C).

(b) Place the first finger of the left hand in the cold water and the first finger of the right hand in the hot water. Leave both fingers immersed for at least one minute.

(c) After one minute, remove both fingers from the jars and dip them repeatedly but alternately in the warm water for about a second at a time (Fig. 1). Notice the temperature sensation in each finger.

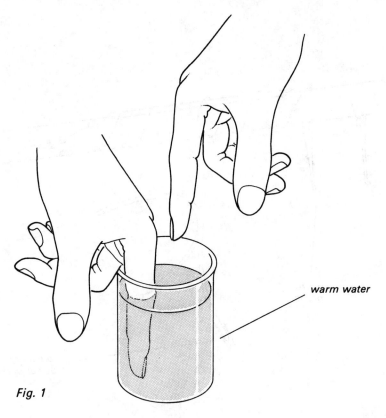

warm water

Fig. 1

Experiment 5.   Discussion

1 What impression did (i) the left finger, (ii) the right finger give about the temperature of the warm water?

2 Why should there be any difference in the sensory information from the two fingers? How could you modify the experiment to test your suggestion?

3 Does the result mean that the skin of your fingers is incapable of judging whether an object is hot or cold?

4 What does the result suggest about the way in which the skin responds to temperature?

Experiment 6. Location of stimuli

(a) Place a glass marble on a non-slippery surface, e.g. on the page of an open notebook.
(b) Cross the first and second fingers of one hand and press on the marble with the tips of these fingers (Fig. 2).
(c) Close your eyes and roll the marble firmly, forwards and backwards, side to side and then with a circular motion for about 30 seconds.
(d) Notice any unusual impressions you receive from the fingers.

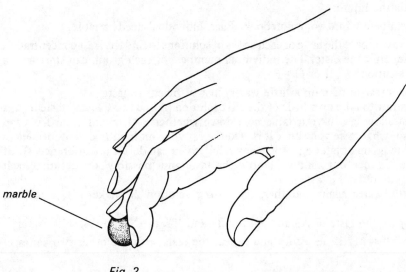

marble

Fig. 2

Experiment 6. Discussion

1 Was there anything unusual in the sensory impressions received from the crossed fingers?
2 Offer a simple explanation of any such impression.

Experiment 7.   Sensitivity to taste

(a) Three solutions are provided which may be sweet, sour, bitter or salt. Each solution has to be diluted to make a range of increasing concentrations. Each group should make up one series of dilutions, A, B or C. These can be exchanged with other groups so that all three series of solutions are tasted.

(b) Label 4 test-tubes A1, A2, A3 and A4 (or B1-B4, or C1-C4).

(c) Use a graduated pipette to place 1 cm³ of the solution in tube 1, 2 cm³ in tube 2, 3 cm³ in tube 3 and 4 cm³ in tube 4. Collect a beaker of clean tap water.
Use the graduated pipette to add tap water to each to make the volume up to 15 cm³ (i.e. 14 cm³ to tube 1, 13 cm³ to tube 2 etc., Fig. 1). Close each tube in turn with your thumb and invert it several times to mix the liquids.

(d) Copy the table given below into your notebook. Each individual needs a table.

(e) You are now going to compare the taste of each series of solutions to find the least concentration at which you can identify the taste. The individual members of each group can start with a different series of solutions A, B or C.

  (i) Use a drinking straw to take up a little water, just sufficient to taste.

  (ii) Use the same straw to take up a little of the first solution (A1, B1 or C1) and taste it (it is quite safe to swallow it). In your table note down whether it tastes different from tap water and, if so, what taste you think it is. Do not say this out loud in case it influences your partners' judgement later on. Write down 'N' if you can detect no difference, 'D' if you detect a difference but cannot identify the taste and 'possibly' or 'definitely salt etc.' if you can identify the taste.

  (iii) Take up a little water again and then try tasting solution 2. Make a note of your impression.

  (iv) Similarly compare the taste of A3 and A4 (or B3 and B4, or C3 and C4) with water.

(f) Rinse the mouth with water if necessary and repeat the tests with the other two series of solutions.

(g) Make a note of the *least* concentration at which you were sure of the taste and find out from your teacher (i) what solutions were used and (ii) what the concentrations were. Write this information in your table.

Solutions	1	2	3	4
A				
B				
C				

Experiment 7.   Discussion

1 Was the lowest concentration at which you could identify the taste the same for any of the series, A, B or C?
2 If it was the same for any two or three of the substances, does this mean that your tongue is equally sensitive to the tastes, sweet, salt and sour?
3 Try and work out the concentration of sugar you normally have in your tea or coffee.

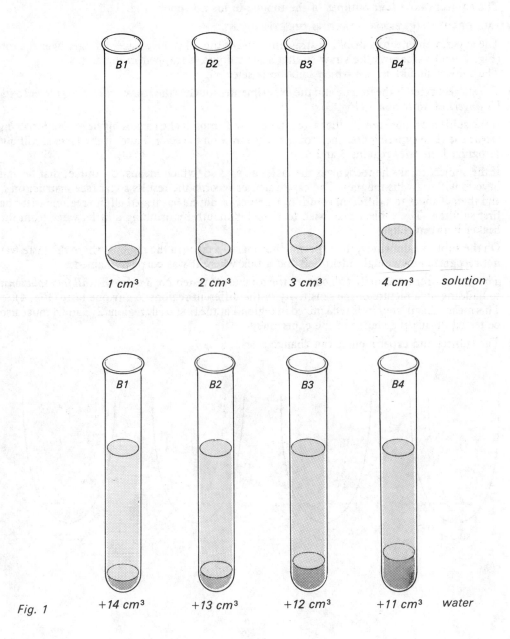

Fig. 1

Experiment 8.   Sensitivity of the tongue to different tastes

The object is to see if some regions of the tongue are better able than others to identify particular tastes.

Work in teams of two, a subject and an experimenter.

(a) The experimenter collects 2 cm depth of each solution in test-tubes labelled 'sweet', 'sour', 'salt' and 'bitter' and places a drinking straw in each tube. He also collects a beaker of clean water.

The subject draws four outlines of the tongue in his notebook (Fig. 1).

READ ALL OF **(b)** TO **(g)** BEFORE STARTING THE EXPERIMENT.

(b) The experimenter dips a drinking straw into one of the solutions, places a finger over the top (Fig. 2) and withdraws the straw so that a little liquid is trapped in the end.

The subject should *not* see which solution is selected.

(c) The subject extends the tongue and the experimenter touches the straw *(still closing the end with his finger)* on to region 1 (Fig. 3).

(d) If the subject cannot identify the taste after about 3 seconds, he shakes his head, *but leaves his tongue out*. The experimenter then touches the straw on to region 2 and, if the taste is still not recognized, he tries regions 3 and 4.

(e) If the subject thinks he recognizes the taste, he says so (which means, of course, that he will have to withdraw his tongue). The experimenter records the results so far (see instruction **f**) and then changes to a different solution, whether or not he has tested all the regions with the first solution. The subject may wish to rinse his mouth by drinking a little water from the beaker between tests.

(f) On the subject's drawings, the experimenter marks a cross in the regions where the taste was not recognized or wrongly identified and a tick where it was correctly named.

(g) The tests are repeated so that *all parts* of the tongue are tested *2 or 3 times* with *all four solutions*, so building up a picture of the sensitivity of the different regions to any one taste (Fig. 4). This means that if 'sour' was recognized in region 1 at the first trial, regions 2, 3 and 4 must also be tested during the course of the experiment.

(h) The subject and experimenter can change roles.

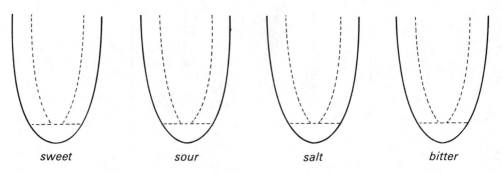

sweet   sour   salt   bitter

Fig. 1

Experiment 8. Discussion

1 Which regions of your tongue were most sensitive to each of the four tastes?
2 Suggest an explanation for the differences in sensitivity.
3 How does the method of tasting used in this experiment differ from the way you normally taste liquids?

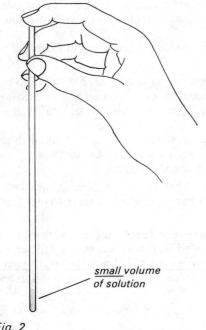

small volume of solution

Fig. 2

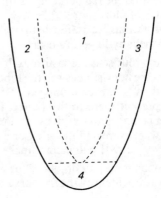

Fig. 3

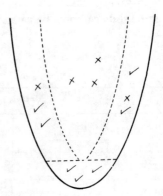

Fig. 4

Experiment 9.   Taste and flavour

You are provided with pairs of solutions. The members of each pair have similar tastes but different flavours.

(a) Label six test-tubes A1, A2; B1, B2; C1, C2.

(b) Collect about 10 cm depth of the appropriate solutions in the tubes and place a drinking straw in each. Also collect a beaker of clean water for rinsing your mouth.

(c) Copy the table given below into your notebook.

(d) Use the straws to taste each of the pairs in turn so that you can identify and name them correctly. Solutions A both taste salt, but one has a garlic flavour. Solutions B taste sweet but have different flavours, e.g. rum or almond. Solutions C should be tasteless but one has a flavour of cloves.
The solutions are quite safe to swallow but you may wish to rinse your mouth from time to time to dispel any lingering tastes.

(e) To test your sense of taste and flavour, take the first pair of test-tubes, hold them so that the labels are not visible, swap the tubes about in your hand so that you do not know which is which and then taste each of the pair. (If you are working in pairs one person can 'shuffle' the tubes and present them to the subject for tasting.)
After tasting both solutions, decide what the two flavours are and then look at the labels to see if you were correct. Put a tick or a cross in the appropriate column of your table.
Do not guess. If you cannot make up your mind, score it as 'wrong'.

(f) Keep the same pair of tubes, but swap them around as before while concealing the labels, taste both again, identify the flavours and record whether you were right or wrong. Repeat this twice more.

(g) Rinse your mouth with water to dispel any lingering taste and repeat the tests with the *same* pair of solutions but this time pinch your nostrils firmly while you are drawing up the liquid and tasting it. *Make your decision about flavour with your nostrils still closed.* Record in your table whether you were right or wrong. *Do not guess.* If you cannot make up your mind, score it as 'wrong'.

(h) Repeat the test three more times with your nostrils closed each time and record your successes and failures.

(i) The same tests can now be carried out for solutions B and C.

Trials	Nostrils open				Nostrils closed			
	1	2	3	4	1	2	3	4
A1 salt A2 garlic salt								
B1 sweet/rum B2 sweet/almond								
C1 water C2 cloves								

Experiment 9.   Discussion

1 Was there any difference in the number of successful identifications of flavour with your nostrils (a) open and (b) closed?
2 Was there any difficulty in recognizing the taste of the solutions with your nostrils open or closed?
3 What do the results suggest about the ability to recognize taste and flavour?

Taste each of the pair in turn

Experiment 10.   Judgement of distance

The object is to compare your ability to judge distance when using both eyes or only one eye. Work in teams of two, an experimenter and a subject.

(a) The experimenter, without allowing the subject to watch, sticks the coloured pins upright into the marked block. Each pin is placed at an intersection of the lines so that no two pins are on the same line whether the block is viewed from the side or the front (Fig. 1). Avoid making a regular pattern as far as possible.

(b) The subject closes one eye and keeps it closed.

(c) The experimenter passes the block to the subject, tilting the block away from the subject so that he cannot see the marked grid. The subject holds the block quite steady about 30 cm away, tilted so that he can see the front edge of the block and the pins but *not* the marked grid on the top of the block or either of its sides.

(d) The subject now calls out the colours of the pins, in order from the *front* to the back and the experimenter writes this down.

(e) Still holding the block in the same position, the subject opens both eyes and calls out the order of the pins again from front to back.

(f) The correct answers are marked.

(g) The experimenter changes the position of the pins and the experiment is repeated twice more.

(h) The average number of correct answers for one eye and both eyes are compared.

(i) The subject and experimenter change roles.

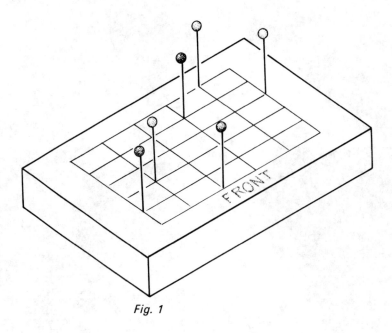

Fig. 1

Experiment 11.   Inversion of the image on the retina

(a) Hold the apparatus close to your eye, with the card touching your face. Point it towards a brightly lit area or window and look at the pin-head through the pin-hole. Describe what you see.

(b) Again hold the apparatus close to your face but this time with the pin nearer to you than the card. Hold the wooden base against your cheek-bone and move the apparatus about slightly while looking through the pin-hole until the pin-head can be seen as a silhouette against the pin-hole. Describe what you see.

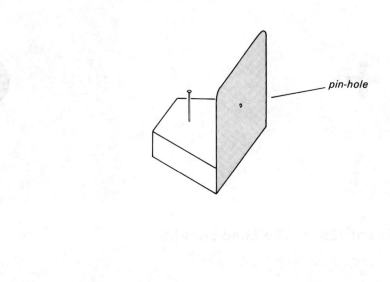

pin-hole

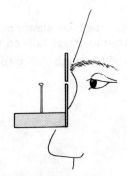

Firstly view the pin through the pin-hole

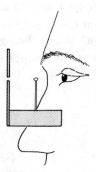

Then line up the pin in front of the pin-hole

Experiment 12a.   The blind spot (1)

(a) Close your left eye.

(b) Hold the book about 60 cm away from your face.

(c) Concentrate on the cross with the right eye and slowly bring the book closer to the face, still concentrating on the cross.

(d) Repeat the experiment with the book held sideways so that the dot is above the cross, and also with the book upside down with the dot on the left of the cross.

(e) Describe what you see in each case and explain it as far as possible.

Experiment 12b.   The blind spot (2)

(a) Close the left eye and hold the book at arm's length.

(b) With the right eye, concentrate on the cross and bring the book slowly nearer your face. At first, the short line on the right will disappear because its image falls on the blind spot.

(c) When the short line reappears, the image of the gap will be on the blind spot. Describe what you see at this point.

Experiment 13.   Duration of visual impressions

(a) Place the book flat on the bench or table.
(b) Close one eye and concentrate on the dot in Fig. 1A for about 10 seconds.
(c) Repeat instruction (b) but with Fig. 1B.
(d) Record your impressions in each case.

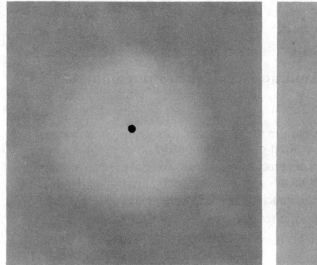

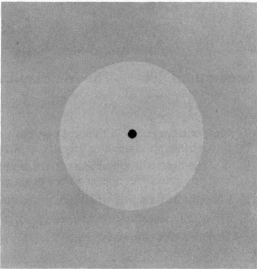

Fig. 1A   Fig. 1B   T. M. Cornsweet

Experiment 14.   Retinal capillaries

(a) Close one eye and hold the card close to the other.
(b) Look through the pin-hole at a brightly lit sheet of plain white paper about 30 to 40 cm away.
(c) Move the card about very slightly with a circular motion so that you can see through the pin-hole all the time.
(d) Allow your eyes to relax and a net-like pattern of capillaries will appear against the white background after a few seconds provided you keep the card moving.
(e) There are no capillaries over the fovea. Can you pick out this area in the image you are forming?

Experiment 15a.   The iris diaphragm (1)

For this experiment the room must be darkened but not blacked out.

(a) If you are working in pairs, sit facing each other and take it in turns to hold a bench lamp or torch about 10 cm from the *side* of your partner's face.

(b) Switch it on and watch the pupil of the eye very carefully.

(c) Switch the light on and off at 3 or 4 second intervals.

If you are working on your own, you must look closely at a mirror while switching the light on and off. Study the pupil of the illuminated eye.

Experiment 15b.   The iris diaphragm (2) (Broca's pupillometer)

(a) Close one eye and hold the card very close to the other, actually touching your face, and look through the pin-holes at a brightly lit sheet of plain white paper.

(b) The pin-holes will appear as a pattern of unfocused light discs. Some of the discs will appear to overlap, others will appear to be separated.

(c) Concentrate on a pair which are nearly touching but not overlapping.

(d) While still concentrating on this pair of discs with one eye, open the other.

(e) Describe what you observe.

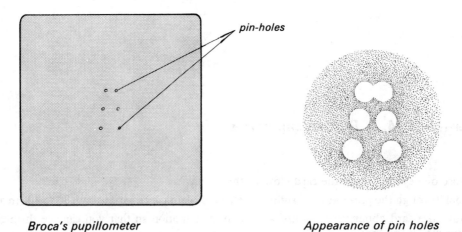

Broca's pupillometer   Appearance of pin holes

Experiment 16a.   Binocular vision: eye dominance

(a) Keep both eyes open and hold a pen or pencil upright at arm's length.
(b) Quickly move the pencil to come exactly in line with a more distant vertical object, such as a window frame or lamp-post.
(c) Close and open the left eye.
(d) Close and open the right eye.
(e) Note any change in the apparent position of the pencil and whether it was the closure of the left or right eye which produced it.
(f) Which eye did you use in lining up the two objects?

En garde!

Experiment 16b.   Binocular vision: double vision

(a) Keep both eyes open and concentrate on the square drawn below.
(b) Rest a finger lightly against the upper eyelid at the outer corner of your less dominant eye (experiment 16a) and press very gently.
(c) The two images formed by the eyes cannot now be properly correlated by the brain, and double vision results.

Experiment 17a.   Colour vision: recognition of colour

It is necessary to work in pairs, a subject and an experimenter.

(a) The experimenter selects a number of coloured pencils (preferably including red, blue, green and yellow) which are similar in all respects other than colour. He also draws in his notebook, a table similar to that shown on p. 283.

(b) The subject fixes his gaze on an object straight ahead and extends his right arm sideways at right angles to his line of forward vision. The experimenter holds the cardboard 'protractor' under the subject's arm (Fig. 1).

(c) The experimenter places one of the coloured pencils in the subject's right hand and the subject, without turning his head or moving his eyes, tries to say what colour it is.

(d) If the subject cannot distinguish the colour or describes it wrongly, he then moves his extended arm slowly forward over the card, still directing his gaze straight ahead, until he can correctly recognize the colour of the pencil.

(e) The experimenter marks in the table the angle made by the front of the arm where the colour was first accurately identified.

(f) The experiment is repeated with other colours, but some colours should be used twice or more to see if the results are consistent.

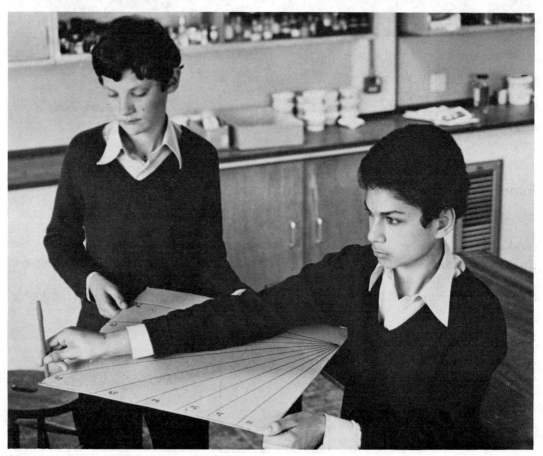

Fig. 1

Experiment 17a.   Discussion

1 Is there an angle from the mid-line beyond which no colour can be recognized?
2 Are some colours more easily recognized than others on the periphery of the retina (i.e. not the fovea)?
3 In the retina, only the receptors called cones are able to distinguish colour. What conclusion do you reach about the distribution of cones in the retina?

Colour	Angle		

Experiment 17b.   Colour vision: complementary colours

(a) Stare hard at the coloured square for 10 seconds and then transfer your gaze to the X for 5 seconds.
(b) Describe what you see.

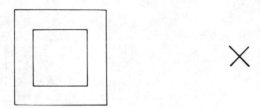

Experiment 18.   Eye and hand co-ordination

It is best to work in pairs; a subject to conduct the trials and an observer to time them. It is possible, however, to do the experiment on your own.

(a) Copy the table given below into your notebook.

(b) Use adhesive tape to stick the sheet with the 'star' outline to the bench, with the MIRROR line away from you and not less than about 30 cm from the front edge of the bench.

(c) Place the mirror vertically on the line and support it, if necessary, with a clamp or a book placed behind it.

(d) Clamp a piece of card in a clamp stand and arrange it between yourself and the mirror (see Fig. 1) so that when you put a pencil on the 'start here' circle (looking in the mirror), you can see the star and your hand in the mirror, but not by direct vision.

(e) Use the right hand if right-handed, or left hand if left-handed, and place the pencil point on the 'start here' circle.

(f) The observer notes the time or starts the stop-clock and the subject, moving in the direction indicated by the arrow, traces round the outline of the star, *looking only at the mirror image* and making the line, however erratic, pass through *every circle*. When the outline is completed, the observer notes the time taken and enters it in the table.

(g) The pencil line is rubbed out, or the sheet replaced, and the *same subject* repeats the trial but with the opposite hand, and the observer notes and records the time as before.

(h) The trials are repeated at least twice more for each hand, running the trials consecutively, i.e. 2 trials with the right, followed by 2 trials with the left, and the times noted and recorded.

Time to complete the drawing	
Right	Left

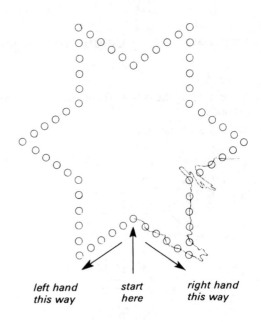

left hand
this way

start
here

right hand
this way

The line must pass through each circle
no matter how erratic it is

Experiment 18.   Discussion

1 Considering only the times for the *first* trial of each hand, does there appear to be a significant difference between them?

2 Suggest possible explanations for any differences.

3 As the trials proceeded, was there a decrease in the time taken to complete them? If so, was there a difference in the rate of improvement for the different hands?

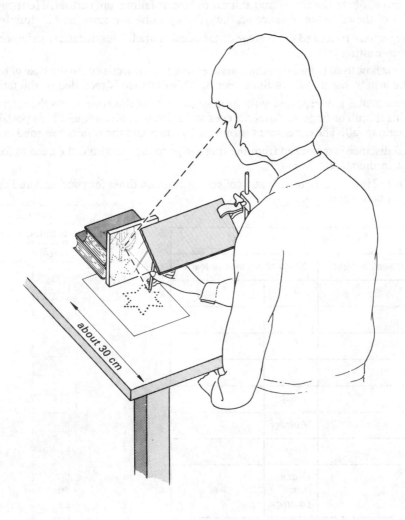

Fig. 1

Experiment 19.   Reaction time

It is necessary to work in pairs for this experiment.

(a) Copy Table 1 into your notebook

(b) The subject marks a pencil line down the centre of his thumb-nail and sits sideways at a bench or table with his forearm resting flat on the bench and his hand over the edge (Fig. 1).

(c) The experimenter holds a ruler vertically between the subject's first finger and thumb with the zero opposite the line on the thumb but not quite touching either the thumb or fingers.

(d) The subject watches the zero mark and, as soon as the experimenter releases the ruler, the subject grips it between finger and thumb to stop it falling any further. He then records in column 1 of the table the distance on the ruler opposite the mark on his thumb.

(e) This is repeated 4 times and the average distance calculated. This distance can be converted to a time by consulting table 3.

(f) The ruler is now used to measure the distance from the subject's eye to the base of his neck and along the arm to the middle of the forearm. This distance is recorded in the table.

(g) The experiment is now repeated with the *same* subject but this time he lets the ruler rest lightly against his thumb or fingers, closes his eyes and grips the ruler as quickly as possible after he *feels* it begin to fall. The experiment is repeated 4 times and the results recorded in column 2.

(h) The total distance is measured from the finger-tips to the head, and the head to forearm, and recorded in the table.

(i) Copy Table 2 into your notebook and collect the reaction times for your class and calculate the average in both cases.

Table 1			
Speed of response to sight		Speed of response to touch	
1		1	
2		2	
3		3	
4		4	
Total		Total	
Average		Average	
Time		Time	
Distance (nerve pathway)		Distance (nerve pathway)	

Table 2			
Reaction times: class results			
sight		touch	
Total		Total	
Average		Average	

Experiment 19.   Discussion

1 Was there a significant difference in reaction times using sight and touch (a) for your own results, (b) for the class results when averaged?

2 The nervous pathway for the motor impulses from brain to forearm was probably the same for both experiments but the sensory pathway from eye to brain was much shorter than from finger-tips to brain. What differences in results would this lead you to expect?

3 State whether you think the class results fulfilled this expectation. If they did not, discuss possible reasons for their not doing so.

4 Assuming, for the moment, that in the touch experiment, the time of reaction is the same as the time taken for the nerve impulse to travel from the sensory organs in the finger-tips to the brain and back to the forearm, calculate the speed of conduction, in cm per second, of the nerve impulse by dividing the distance travelled by the reaction time.

5 The conduction velocity of nerve impulses in mammalian nerve fibres ranges from 2 to 100 m per second. Your results are probably about 5 to 15 m per second. What are the faults in the assumption made in question 4 that might account for this low value?

6 Although it was the fingers which gripped the ruler, you were told to measure the distance to the forearm. Why was this?

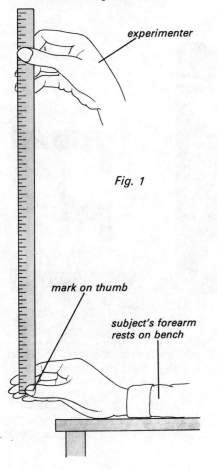

experimenter

Fig. 1

mark on thumb

subject's forearm rests on bench

distance in cm	time in seconds		distance in cm	time in seconds
1	0.045		21	0.207
2	0.064		22	0.212
3	0.078		23	0.217
4	0.090		24	0.221
5	0.101		25	0.226
6	0.111		26	0.230
7	0.120		27	0.235
8	0.128		28	0.239
9	0.136		29	0.243
10	0.143		30	0.247
11	0.150		31	0.252
12	0.156		32	0.256
13	0.163		33	0.260
14	0.169		34	0.263
15	0.175		35	0.267
16	0.181		36	0.271
17	0.186		37	0.275
18	0.192		38	0.278
19	0.197		39	0.282
20	0.202		40	0.286

Table 3

Experiment 20.   Perception

The exercises in this book have been mainly concerned with the functions and limitations of the senses of touch, taste and sight. There is, however, a vast difference between the information sent to the brain by the sense organs and the interpretation made by the brain. This interpretation is called perception and depends, to a great extent, on learning and past experience.

For example, Fig. 1 is a two-dimensional pattern of straight lines and its image on the retina is the same, but it is perceived by most people as a cube. Moreover, it can be seen as a cube with the shaded area either as its face or the inside of the back wall. The image on the retina does not change but the perception does. The brain does not settle for one or other view because both are equally acceptable (probable).

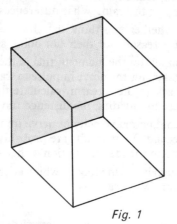

Fig. 1

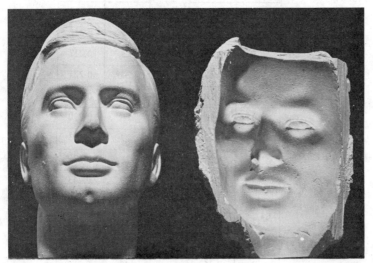

Fig. 2

R. L. Gregory

Fig. 3

Fig. 2 shows two models of the human face but the one on the right is a mould, i.e. it is concave. Both are illuminated in the same way but the right hand one is perceived as a normal face. The brain refuses to interpret the sensory information as a concave face, perhaps because such an object is contrary to all previous experience (i.e. improbable).

In Fig. 2 the brain refuses to 'see' an unlikely object, a hollow face, but in Fig. 3 it is prepared to 'see' an object which is not there, namely a solid letter 'E'. Although the latter is not drawn, the brain 'sees' it as the most probable explanation of the three, otherwise meaningless, angular black areas.